LIVE TV FROM THE MOON

By Dwight Steven-Boniecki

An Apogee Books Publication

Live TV From The Moon
ISBN 9781-926592-16-9 - ISSN 1496-6921
©2010 Apogee Books/Dwight Steven-Boniecki

All rights reserved under article two of the Berne Copyright Convention (1971).
Published by Apogee Books an imprint of Collector's Guide Publishing Inc., Box 62034, Burlington, Ontario, Canada, L7R 4K2
http://www.apogeespacebooks.com
Cover: Robert Godwin/ Photograph of the antique Bush TV-12 television set courtesy if the Early Television Museum, Hilliard, Ohio
Printed and bound in the USA

For my Father, who passed his enthusiasm for the technical achievements of Apollo on to me, and for all the people who worked hard to build the TV systems which brought live video from the lunar surface into our living-rooms.

CONTENTS

Foreword by Stan Lebar	7
Acknowledgements	9
Introduction	11
Chapter 1. The Early Decisions	13
Chapter 2. Faith 7 Slow Scan Television	27
Chapter 3. RCA and the Block I TV Camera	37
Chapter 4. AS-204 - Apollo 1 TV Plans	47
Chapter 5. Apollo 7	55
Chapter 6. Apollo 8	67
Chapter 7. Apollo 9	79
Chapter 8. Color Television	91
Chapter 9. Apollo 10	105
Chapter 10. Apollo 11	113
Chapter 11. Apollo 12	161
Chapter 12. Apollo 13	171
Chapter 13. Apollo 14	179
Chapter 14. Apollo 15	187
Chapter 15. Apollo 16	211
Chapter 16. Apollo 17	221
Epilogue	233
Bibliography	235

In Memory of Stan Lebar

FOREWORD

Dwight Steven-Boniecki has written a book about the trials and tribulations of what was involved in bringing the Apollo television to the home television sets throughout the world. Originally the television cameras that were carried on-board the spacecraft were defined by NASA as Non Mission Critical, meaning that whether the television camera worked or not would not have any effect on the success or failure of the mission of going to the moon and returning to earth successfully.

However, to the person who went about their everyday lives in 1969, the Apollo 11 scheduled live broadcast of the first person to step on a celestial body other than earth on July 20, 1969 was not only Mission Critical, it had become "The Mission" once the astronauts stepped onto the moon. The telecast was seen by the largest world audience ever as Neil Armstrong made his way down the ladder and stepped on the surface of the moon.

Mr. Steven-Boniecki has done what no one else has attempted. He has written the details of how all this was accomplished in a manner that the everyday viewer can follow. It's the story of how a small group of talented engineers came together to create the television cameras to record those moments in history. Their work allowed the world to watch the magic of those moments unfold, live as it happened, on television sets in their own living rooms. The many details in his book are based on extensive research and numerous conversations with those who were intimately involved in every phase of the process that created and managed these very special television cameras.

I recommend this book to anyone who ever asked, "How in the world did they ever manage to provide live TV from the moon back in 1969?" He's answered that and a great deal more.

Stan Lebar,

Westinghouse Electric Corporation, Retired

Program Manager Apollo Lunar TV Camera

July 20, 2009

ACKNOWLEDGMENTS

Just like the Apollo missions, this book could not have happened had I not been fortunate enough to be standing on the shoulders of giants. I am indebted to a huge number of people whom I would like to thank here. Each one of them was responsible for helping me see this project through: Colin Mackellar, whose unending generosity allowed me great insight into the history of tracking stations. Stan Lebar for his Westinghouse anecdotes, photographs and cross-Atlantic telephone discussions into the very early hours of my German mornings, and Sam Russell for his RCA anecdotes, photographs and numerous replies to the most intricate and obtuse question I could throw at him.

To John Lowrance for his explanation of the Block I RCA TV camera which unlocked my stalemate in beginning the story of the TV camera contracts. To Stan, Sam and John: the cameras were your babies, and I hope I have done them justice in this book. John Saxon and Mike Dinn for showing me where TV signals first arrived from the moon and for their Honeysuckle Creek anecdotes as told "on-site". Thanks to Ed von Renouard who graciously spent a few hours chatting to me via telephone recounting his moments at Honeysuckle as the television came down the line.

To Glen Nagle at the Tidbinbilla Tracking Station Visitor's Center, thanks for letting me use the ASTP TV camera! Bill Wood for providing the general public with much of the TV documentation scans and fantastic essay which at first disheartened me and then spurred me on, and also for his Goldstone anecdotes – not to mention the 11th hour assistance in color sequence and color converter clarifications.

Thanks to the following astronauts: Walter Cunningham for his TV camera recollections on Apollo 7, Alan Bean for demonstrating first hand how fragile the TV camera was on Apollo 12, Al Worden for his recollections of using the TV camera on Apollo 15. Ed Mitchell for his memories of the ease of use with the Westinghouse camera during his Apollo 14 moonwalks. Jim McDivitt for explaining to me the fight he had to mount to ensure the GCTA flew on the later missions.

To Ed Fendell for sharing his recollections as "Captain Video" during the later Apollo missions. A big thank you to Max Engert for his recollections in NASA's selection of the supplier for the color TV cameras. Huge thanks must go to Jean Grant from the JSC Archive Library for her patience and assistance in locating many of the more obscure NASA documentation. Michele Burton, Debbie Sander, Elisabeth McDonald, Jillian Kelso, Margaret Gwynne, and Anna Juniper, for their assistance at the Jerzy Toeplitz Library of the Australian Film, Television and Radio School for the unlimited access to all the SMPTE journals I could name. Huge thanks also to Jody Russell, Still Photo & Video Librarian Researcher at the Media Resource Center NASA Johnson Space Center for the photos that made my jaw hit the ground!

To Eric Jones and Kevin Glover for

the Apollo Lunar Surface Journal; the first stop for anyone interested in the Apollo missions (and that included me in the early days of this project). Thanks to Calvin Mackey from NASA's Technical Reports Server for the wealth of information pertaining to Apollo-era TV. A big thanks to Mark Gray of Spacecraft Films for putting every available millimeter of Apollo era footage whether it be video or film onto high quality DVD, and his perfect timing of making a TV special about the Apollo Television Systems right when I started typing the first chapters of this book. Thanks as well for the generous permission to use screen grabs from your DVD material. As a great counterpart to this book I urge you to obtain his outstanding Apollo mission sets. To Thilo Elsner stationed at the Bochum Observatory in Germany for showing me how the non-English speaking world listened in on the Apollo missions. Robert Pearlmann who maintains www.collectspace.com a site of collectors that also introduced me to numerous resources I would otherwise not have known about. Aleksandra Claxton for keeping my writings centered on the subject matter and for the painful criticisms which ultimately lead to a better end-product, and for most importantly being the rock through all the highs and lows in seeing this project through to conclusion.

To Oliver Rowley much thanks for his pep talk right at the beginning of this project. A big thank you to Dr El-Baz for his recollections of his proposal for unmanned LRV traverse of the Hadley region. To Doug Bennett for his invaluable critique of the opening chapters--talk about changing water to wine! Thanks to Phil Hoffman for his insight into the electrical switch he designed. Thanks to Don Davis for use of his assembled Apollo 11 panorama. To Christel Jansen for supplying me with a high speed scanner which converted all the hard copy JSC documents into PDF file format. You saved me months of work! To JL Pickering and Ed Hengeveld for help in tracking down photos of the cameras; you two are the masters of photographic archives.

To David Harland and Alan Lawrie for their votes of confidence and insider tips on getting this epoch published. Thanks to Mido Fayad and Ida Shahnazari for allowing me to work on this project along side my regular TV transmission controller's job. Thank you to Pamela May, who supplied some of the lesser known TV engineering photographs.

Thanks to Rob Godwin and Matthew Heimbecker for believing in this project based solely on an introductory email and for subsequently publishing it. Lastly, it is rarely mentioned when discussing spaceflight, but to all the inmates of Dora Mittelbau concentration camp: your forced contributions under shocking and atrocious conditions to the development of spaceflight will never be forgotten. It is a sad irony that humanity's most crowning achievement was born in part out of its most disgraceful chapter.

THE FAMILY Handyman
thefamilyhandyman.com

FHM5AX

Save $19.92!

Big Holiday Gift Savings with 50% off!

YES! I want to send 1 year of The Family Handyman to someone special at your low gift rate! Give The Family Handyman – only $19.98 for 1 year. That's 50% off the cover price. Send no money now – we'll bill you later.

11/1998

SEND MY 1 YEAR GIFT TO:

Name _____

Address _____

City _____ State _____ Zip _____

BILL ME:

Name _____

Address _____

City _____ State _____ Zip _____

Each 1-year (11 issue) subscription includes a special issue, which counts as 2 issues in a subscription. Subscriptions outside the US are $29.98. Offer expires June 2014.

NO POSTAGE
NECESSARY
IF MAILED
IN THE
UNITED STATES

BUSINESS REPLY MAIL
FIRST-CLASS MAIL   PERMIT NO. 350   HARLAN IA

POSTAGE WILL BE PAID BY ADDRESSEE

THE FAMILY
Handyman
PO BOX 6133
HARLAN IA 51593-3633

SAVE 50%!

INTRODUCTION

My earliest recollections of the Apollo missions are of Apollo 17. I was 3 years old and watching the moonwalks when my father took me outside, pointed to the moon and asked if I could see the men walking on it. I wished I could have, but was content to watch the TV pictures of them instead. As a child I longed to see more visits to the moon, but they never happened. More than 40 years after the first historic landing of Apollo 11, I'm still patiently waiting for the next set of missions to explore the moon.

The first time I ever saw the entire Apollo 11 EVA was on the 20th anniversary when the A&E cable network in the USA repeated the telecast in its "As it Happened" series. The excitement was certainly there even though the events occurred two decades earlier. It wasn't until 2003 when I would get to see any of the missions in their entirety thanks to Mark Gray's Spacecraftfilms DVD series. The sets had all the missions and featured all of the footage. I watched, fascinated.

One thing perplexed me as a television engineer. How did the color TV cameras work on the moon? Why did the color look funny any time someone moved across the screen? I embarked on a quest to find out the answers to my questions, and found to my dismay that very little had been written on the topic. This struck me as strange, considering that it was only through television that the general public could truly experience the excitement of the missions. My notes began to increase and I found the more I looked, the more I found. Through discussions with fellow enthusiasts, the astronauts themselves, and the very people who helped create the revolutionary video technology, my research suddenly turned into an extensive library of material. Someone (I'm afraid I don't recall who, exactly) then suggested I do an article about the television cameras for a magazine. There was, however, just too much information for a simple article, and that was the foundation for writing this book.

Locating the relevant data was nothing short of an adventure in itself. Certain documents had gone astray in the four decades since the Apollo missions were first conceived. At the time of Apollo, television was in what many consider its 'Golden Age'. The medium itself had been established on the decades of research beginning in the Nineteenth Century. Using the designs of early TV research, RCA and Westinghouse engineers created a new type of video camera capable of sending color images across the expanse of space.

I obtained notes, conducted interviews, and examined videos and photographs. The book became a quest to have as complete a record of how television found a seat on a rocket to the moon as possible. Unexpectedly, I seemed to have become an expert on the Apollo television systems.

This project was a huge challenge for me. I had to look outside of NASA to find some of the more technical material. I visited locations around the world looking for as much information as I could. In some cases I had to re-learn basic TV theory in order to refresh my memory. Then, of course, translating technical reports into plain English is no easy task.

This books deals predominantly with the Apollo TV systems. Of course, the unmanned probes of Ranger and Surveyor sent to the moon prior to the successful Apollo landing had TV systems, but I have purposely not discussed them here unless they are directly related to the Apollo missions (for example, Apollo 12). Similarly Gemini, while absolutely critical to the success of Apollo is not covered, precisely because it never had any television capability at all during its 10 manned flights.

One point of contention is the color sequence of the color TV cameras. Conflicting accounts exist describing the sequence in which the three primary colors were scanned. After painstaking examination and cross-referencing (including fame-by-frame analysis of the color transmissions I have determined that the correct sequence of colors was Red-Green-Blue rather than Red-Blue-Green, as is sometimes stated.

I sincerely hope that the reading experience is not dampened by the technical nature of the subject. Without a doubt, I could not have finished the book without the overwhelming generosity of Stan Lebar, who headed the Westinghouse lunar camera development team in the 1960's. I had always envisioned him reading the finished product, which detailed a part of his life that he was extremely passionate about. It deeply saddens me to realize that due to his untimely passing on December 23, 2009, he will never get to see the book in the complete form you now hold in your hands.

As of 2010, it is the only written work completely documenting the Apollo television systems.

CHAPTER 1: THE EARLY DECISIONS

"Television won't last. It's a flash in the pan."

Mary Somerville, radio presenter, in 1948.

December 14, 1972. More than 188 hours into the final lunar mission of Apollo 17 on the lunar surface in the Taurus-Littrow valley. Over ten years had passed since television had been proposed on the Apollo lunar missions. Now, as a lone television camera pointed at the stationary Lunar Module "Challenger", earthbound viewers could watch live as two astronauts, Gene Cernan and Harrison Schmitt prepared to launch, and bring to a close mankind's preliminary explorations of the moon. In the completely airless lunar environment devoid of all life, it must have been an odd sight: the abandoned lunar roving vehicle with a golden umbrella-shaped antenna pointing at a bright blue and white planet in the black lunar sky. Several centimetres below this structure, a shoe-box sized device with a lens attached to the front of it, swivelled up, down, and from left to right, as invisible extended hands from Mission Control guided it via remote control. With approximately one minute left Cernan said to his crewmate, "Take your final look at the valley of Taurus-Littrow, except from orbit." The static lunar scenery appeared as a brownish grey with the sole evidence of human presence, a spider-like vehicle, framed neatly in the middle of the shot. Live voice communication continued from the spacecraft and Schmitt counted down to the launch time, "Proceeded 3, 2, 1..." as the camera zoomed out in anticipation of what was about to occur. "Ignition!" called Schmitt as pieces of the gold-colored Mylar insulation protecting the lower stage of the craft flew out from the lunar module in all directions.

The television image appeared to spurt rainbow-like confetti, which was, in fact, an artefact from a small wheel, with red, green and blue filters, spinning inside the camera. With each color separately sent to earth in sequence, special ground equipment was necessary to recombine these separate hues into a full-color image. On regular studio video cameras, the three colors would be processed simultaneously providing fluid, natural-looking movement, however with the lunar camera signal, because color-converters repeated each frame two more times while the next color arrived, fast moving objects would appear as color streaks providing a dynamic rainbow of red, green and blue.

The upper stage of the LM rose rapidly from its launch pad towards the unseen Command Module orbiting the moon 50 or so miles overhead. "We're on our way, Houston!" exclaimed Cernan moments after blasting off from the lunar surface. The camera on the rover held the rising lunar module in shot; a bright speck against a completely black sky. "Pitch over" announced the crew in unison as the craft tilted forward according to the view captured by the television camera on the lunar surface. The orange glow of the rocket engine became visible on the TV images as the spacecraft obliquely headed into orbit.

An artist's concept of the televised launch of the LM. (RCA Astro Electronics Division Photo 75-4037)

Due to the enormous distance travelled by the remote control commands, flight controller Ed Fendell, sitting in Mission Control in Houston, had three seconds earlier pre-empted the flight path of the LM and had exquisitely managed to follow it upwards. After all, he had covered two previous lunar launches, and on this final one, he was able to get it just right and fully live up to the nickname bestowed by his colleagues, "Captain Video".

Only a handful of the many people watching around the world realised that at one point the very idea of television from the lunar surface, or any Apollo spacecraft for that matter, was effectively struck off the mission objectives. Even fewer had any inkling that those who wanted lunar television had to mount a serious fight to guarantee it a place in arguably the most important human exploration of all time.

The camera was operating far beyond what any of the original mission planners had envisioned a decade before. A steady stream of information was being uplinked via the Lunar Rover's Lunar Communications Relay Unit (LCRU) system. This information travelled a quarter of a million miles to several tracking stations on earth. The tracking stations were spread out around the planet to ensure that at any given time during the earth's daily rotation, an uninterrupted view of the moon could be maintained by the network of stations which subsequently received the incoming radio signals from its surface. From there the raw television information was then passed on via cable, microwave signal transmission, and satellite to Houston.

Finally the signal was reassembled into a conventional color television signal via a video disk storage system and sophisticated electronics. It then travelled from NASA to the television networks and subsequently into people's homes around the world - with only 12 seconds passing since the signal had left the lunar surface!

Like everything in the Apollo Lunar Program, the television camera had its origins in lengthy and detailed studies and design proposals. One of the most common misconceptions regarding the entire Apollo project is that its genesis was when President Kennedy gave his famous speech to Congress on May 25, 1961. While his inspirational address is attributed to solidifying the United States' commitment to a lunar program, in reality many aspects of mission planning were already well under way before Kennedy's speech. When the House Select Committee on Astronautics and Space Exploration made public a document entitled "The Next Ten Years in Space" in early 1959, a two-fold goal was notably evident: a reconnaissance mission to the moon, followed by a lunar landing, both to be achieved within one decade. The idea was bold. While the United States had not yet presented a single manned spaceflight, people within a fledgling NASA and the aerospace community already had their sights set on the moon. Wernher von Braun, the German rocket expert who had worked on the Nazi rocket offences during World War Two, had already informed Thomas Keith Glennan, NASA's first director, in December of 1958, how a moon landing may be carried out. The direct-ascent method which he outlined was at that time acknowledged as the most uncomplicated method for landing on the lunar surface. Essentially, such a mission used a gigantic rocket which launched from earth, subsequently landed on the moon and returned again to earth. Using this approach, mission planners at NASA had already begun designing advanced manned spaceflight. The Lunar Orbit Rendezvous method which was ultimately used for Apollo, in which two spacecraft travel together into orbit around the moon, and then the Lunar Module flies to the lunar surface and back again presented unique problems during the mission. However, at the start of 1960 the funding for a project as grand as a lunar landing was anything but guaranteed. The mindset at the time held that any definite lunar mission would occur well and truly after the year 1970.

This way of thinking was, of course, to change dramatically following Kennedy's address to Congress in which he challenged,

"..I believe that this nation should commit itself to achieving the goal, before this decade is out, of landing a man on the moon and returning him safely to the earth. No single space project in this period will be more impressive to mankind or more important for the long-range exploration of space; and none will be so difficult or expensive to accomplish. We propose to accelerate the development of the appropriate lunar space craft. We propose to develop alternate liquid and solid fuel boosters, much larger than any now being developed, until certain which is superior. We propose additional funds for other engine development and for unmanned explorations - explorations which are particularly important for one purpose which this nation will never overlook: the survival of the man who first makes this daring flight. But in a very real sense, it will not be one man going to the moon - if we make this judgment affirmatively; it will be an entire nation. For all of us must work to put him there."

With that, the United States found itself rising to the challenge of a successful lunar mission within a timeframe of less than ten years.

Early in January of 1960, Abe Silverstein, head of the Office of Spaceflight Programs, had recalled his old school lessons about Greek mythology. His suggestion for the project beautifully encapsulated the spirit of the planned lunar missions. The name of the God Apollo who rode across the sky in his sun chariot became the name of the project to take man to the moon. NASA management, prompted by Silverstein's suggestion, also felt the name appropriate. With funding now provided thanks to Kennedy's passionate appeal, the system was in place to tackle the challenge and to establish the United States as the leader in space exploration. The following July, of the same year, at a NASA-Industry Programs Plan Conference, Hugh Dryden, who had served as the Deputy Administrator for NASA since August 19, 1958 announced to the world that the spacecraft intended to fly to the lunar surface and back to earth would be named "Apollo".

Journeying to the moon during the lunar missions would require techniques and equipment which well-surpassed those used on all earlier manned spaceflight. The outward journey, the Command Module (CM) and Lunar Module (LM) docking and separation procedures (necessary to transfer the crews to the spacecraft while journeying to the moon), the LM descent to the lunar surface, the subsequent lunar surface stay, LM ascent, and final journey to earth were just some of the phases of the mission which would require close monitoring by ground stations. As such, these mission aspects required communication procedures vastly different from anything used before.

A single radio wave was planned to convey all of the Apollo spacecraft's communication requirements. The Unified S-Band (USB) signal was designed to use a sole radio carrier frequency in each direction (ground-to-spacecraft, and spacecraft-to-ground) in order to provide combined tracking and communication with the spacecraft. This system was developed as it was one that offered the best solution to the technical hurdles faced by mission planners, while simultaneously needing only a relatively small amount of new development. Had the range of existing tracking and communication systems used prior to Apollo simply been expanded, there would have been a huge and costly development of radar techniques, in conjunction with a major expansion of the radio equipment. Fortunately, equipment was already in operation which utilized S-Band methods and thus considerably less development was required and substantial capital expenditure was avoided. This also had the additional benefit of largely reducing the pre-requisite tracking and communication equipment necessary on board the spacecraft.

In order to maintain constant contact with the spacecraft at lunar distances, it was crucial for NASA to have a communications network which was equally spaced across the globe, so that regardless of the time of day, or spacecraft position, direct contact with it was always retained. Three major sites were selected for the Apollo network which was spread across the planet.

Situated approximately 140 miles north of Los Angeles the Goldstone tracking station was where deep space communications was born. The upcoming pioneer probes which would travel to the moon needed to be monitored or "tracked" allowing earth based observers to know precisely where the craft was, and to read the data it would transmit back to earth.

Honeysuckle Creek, Australia was located approximately 20 miles from Canberra, the nation's capital. In Madrid, Spain the third tracking station was found, completing the ring around earth where contact with the spacecraft of Apollo could be maintained. For all three sites, it was crucial to isolate the extraneous noise from civilization which could impair the incoming weak signal from space. The bigger the antenna diameter, the better the reception, and Goldstone, with its 210 foot antenna built for the Mariner Mars probes, was adapted for the Apollo tracking schedule.

Madrid and Honeysuckle had 80 foot antennas which, while adequate for signal reception, were not scheduled in the proposed mission timelines for prime reception of planned moonwalks (on the Apollo 11 moonwalk, the radio antenna located at Parkes, Australia was also used for TV reception given its 210 foot size).

Collins Radio, contracted to provide equipment for the astronauts to communicate with earth stations and equipment to track and communicate with the Apollo spacecraft, selected Motorola Incorporated Military Electronics Division to develop and construct the USB system. This technique of radio communication was based on a development of the Jet Propulsion Laboratory (JPL) known as a coherent Doppler and pseudo-random range system. This method of spacecraft ranging, whereby the spacecraft distances from the earth are measured, used a ground transmitting and receiving station working together with a spacecraft transmitter and receiver. A code was phase-modulated (in which the phase of a carrier wave is varied in order to transmit the information contained in the signal) at the ground station and transmitted to the Apollo spacecraft. This received code was then re-transmitted to the ground on another S-band radio wave. The two signals were what is referred to as "phase coherent", meaning they maintained a fixed relationship to each other and subsequently, ground equipment could use this signal relationship to extrapolate an accurate measurement of the spacecraft's distance from earth.

Included in the JPL model was the provision for voice and data channels. The voice and telemetry information transmitted from the spacecraft to the ground stations could be modulated onto relevant radio waves, combined with the ranging signals, and used to phase-modulate (PM) the downlink carrier frequency. Additionally, the transmitter was able to be frequency modulated (FM) for the transmission of standard television signals or additional recorded data.

Modulation is a means employed to

send an information-bearing signal over great expanses. A modulated television signal superimposes this information-bearing signal onto a carrier signal which can be transmitted without problems and is normally a high-frequency sinusoid waveform. The three key parameters of a sine wave are its amplitude or volume, its phase or timing and its frequency or speed, all of which can be modified in accordance with a low frequency information signal to obtain the desired modulated signal. It was thus decided that any prospective TV signal would be contained within the S-Band carrier, albeit severely limited at only 500 kHz against the standard TV industry 4MHz bandwidth constraint.

The benefits of television as opposed to film or standard photographic technology was that it was immediate and it was an electronic medium, meaning it could easily be transmitted from virtually any environment or location imaginable. NASA planners, mission controllers and even the general public could theoretically watch in real-time as the lunar explorers conducted their exploration.

However, despite being granted bandwidth allocation, the use of television on the Apollo Project was barely considered in the early years and it was added on later almost as an afterthought; something which might possibly have been of some use for an as yet undefined purpose. In a 1960 set of guidelines listed by Robert G. Chilton the suggestion was made that television may be desirable for the lunar missions. No further discussion was made of what such "desirability" would entail, although the existing notion of direct approach lunar landings would indicate this included assistance in landing the rocket in a vertical position. Earth Orbit Rendezvous would have required the crews on their backs and subsequently requiring the use of mirrors and/or television monitors to see the lunar surface upon settling the spacecraft on the moon.

Chilton had been a B17 Bomber pilot during the Second World War, becoming part of NACA afterwards. This subsequently led him to become a member of the newly formed Space Task Group (STG) for whom he conducted evaluations of contract proposals. Armed with a bachelor's and master's degree earned at Massachusetts Institute of Technology, he was crucial in developing early spacecraft specifications for Mercury, Gemini, and Apollo. The earliest of lunar mission definitions were the direct result of exhaustive reviews conducted by the STG. While initial evaluations concentrated mainly on getting into lunar orbit and then returning home, solid state electronics were deemed essential to the ultimate success of the grand Apollo project. What is apparent in early studies and recommendations by the STG is that video signals could be incorporated into the S-Band signal and thus some provision should be made to allow for the possibility should the need arise. The recommendation by Chilton for television would be a theme returned to long after his initial proposal, and would often times be the cause of much controversy within NASA.

Outside of any mission critical use, television requirements in the initial planning phase were proving to be a hot potato. One faction within NASA, headed by the Public Relations Department saw the use of TV as a huge benefit to the

U.S. space program. Their argument was based on the notion that unlike the Soviet program, the Americans had always maintained a completely transparent operation and therefore the taxpayers who had paid for the lunar missions ought to be entitled to a front row seat as the operations took place. The other group opposed to TV integration strongly objected to the addition of anything which did not directly contribute to the success or safety of the missions. Astronauts were on board to fly the spacecraft and not to provide a "variety show" to earth-bound audiences.

Making matters a little more complicated, the Apollo Technical Liaison Group for Instrumentation and Communications simultaneously drafted a set of guidelines which nearly stopped television development for Apollo dead in its tracks as early as April 1961. They argued that full-resolution high-quality television would swamp all other communication resources, while also adding a bandwidth requirement which was not otherwise necessary. Citing these reasons, the group requested all interested parties to restate their justification for inclusion of the potentially cumbersome television system into mission plans.

Highlighting the perceived threat of the Soviet advances in spaceflight, the STG referenced close-up photography of the moon by the recent unmanned Russian Luna probes as an argument for television. Referring to the desirability of observing the astronauts on the lunar surface, and being able to collect visual data in real-time, a case was made which utilized a more compact television system which would operate well within the resource-limited confines being put together for the manned lunar missions. Moves were already made, as detailed by the STG, to reduce the resource burden a full-rate TV system would place on the spacecraft. This meant that any notion of using bulky, studio-sized cameras was completely out of the question. New technology, which in most cases had not yet been invented, would be required to send video images over the great distances from the moon to the earth and allow them to be viewed on a normal television set, whether it was mounted in NASA's Mission Control, or in the living-room of an ordinary person's home.

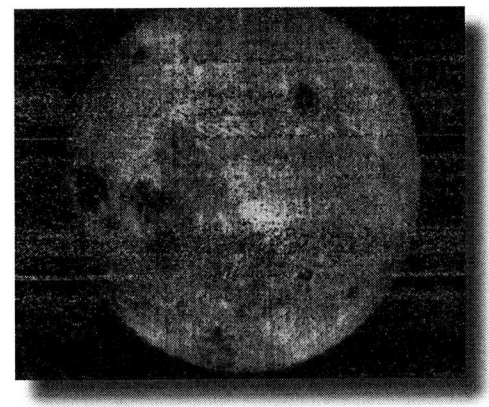

The first image of the lunar far side obtained by the Soviet Luna 3 probe October 7, 1959.

Television, in its analog video signal form is essentially an electrical voltage which varies over time. The television system used in the United States, NTSC (National Television Systems Committee) is made up of 525 vertical lines of resolution. These lines are made via an electron beam which "writes" onto a phosphorous picture tube one line at a time. During a blanking period the scanning beam begins its writing of new visual information onto the screen.

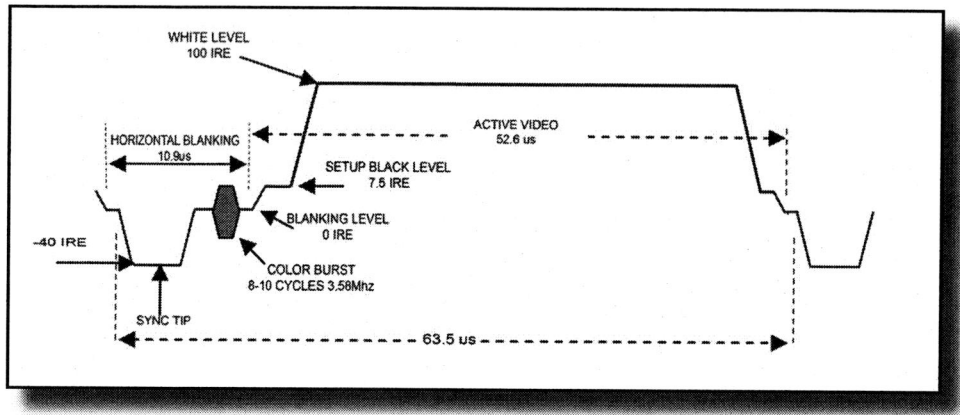

A standard Television waveform.

Each new change of images is made 29.97 times per second. The television scanning beam starts at the top left-hand side of the screen and writes one horizontal line of information. At the end of the line, the beam turns off, moves down, switches back on, and writes the next line. This procedure is repeated 525 times per frame to produce a full picture. The physical makeup of the screen's surface allows it to glow for the duration of the beam's writing over the full 525 lines, and thus a complete picture is seen. The perception of motion is seen due to a phenomenon known as "persistence of vision" whereby the human retina retains an image longer than it is actually presented in reality.

In order to reduce image flickering, a procedure known as "interlacing" divides one frame into two fields each of which contains half the vertical information. The first field comprises of all the odd numbered lines (numbered 1, 3, 5 and so on) while the second field comprises all the even line numbers (2, 4, 6 and so on). After field one is scanned, the beam returns to the top of the screen and writes the second field's information. In effect, each second of motion is made up of 60 separate image scans, thereby achieving the desired result of fluid, flicker-free movement.

The video image in an NTSC signal is comprised of seven sets of electronic information. The waveform includes a horizontal line synchronization pulse, a color reference burst, reference black level, luminance information, color saturation information, color hue information, and a vertical synchronization pulse.

The horizontal line sync pulse directs the scanning beam to a fixed horizontal position. This ensures the line is begun at the same start position for each new line. An additional horizontal blanking period occurs at the end of each written line, while the beam moves back to the new line's start position. By turning the beam off, no extraneous information is written onto the screen during this movement. A similar signal, the vertical synchronization pulse, controls the amount of time the beam switches off while it moves to the starting point of a new line and field.

Color information is handled by both a 3.58Mhz color reference burst, and a 3.58Mhz color subcarrier. Each line of information has this burst before the visual information. It is made up of an eight or nine cycle sine wave, with its phase set at zero. The color subcarrier information controls the saturation and the hue of the colors by being compared to the reference signal.

Picture brightness, or luminance, ranges from a black level of 7.5 IEEE (Institute of Electrical and Electronics Engineers) units, to the highest value of 100 IEEE units for pure white. Corresponding readings within these two settings make up a complete range of greyscale information.

Not all information in a television picture is essential for acceptable performance however, so proposals were considered to streamline the TV signal from the spacecraft. Several methods to trim down redundant television information were put forward in 1961 which made use of techniques which became the benchmark for the system which was later used on the first lunar landing. It was suggested that the vertical line resolution of the video image could be reduced to 350 lines and transmitted at a frame rate of 20 frames per second. In conjunction with planned digital compression techniques it was demonstrated, in theory at least, that a good quality real-time TV picture could be sent to ground tracking stations within the allocated 500 MHz bandwidth.

A January 23, 1962 memo from the NASA Electrical Systems Branch put forward its thoughts on a television system for use on Apollo. In particular it cited the most ideal means of transmitting the signal to and from the spacecraft as being the Delta Modulation method of analogue-to-digital conversion of the TV signal, which is the type of modulation primarily used in voice transmissions where quality is not deemed the essential factor, and where only changes in the signal are sent thus eliminating a substantial amount of unnecessary information. Given that a television signal was inherently a continuous and linked string of information, Delta Modulation was suited to the planned narrow dynamic range of the image expected during lunar missions without the dreaded fall in picture quality. Because of this, a digital television system was generally accepted as being ideal due to the compression that could be applied to the resultant signal, which would reduce transmitter power and bandwidth. Not surprisingly, investigations and comparisons were conducted to establish the trade-off issues for both analogue and digital television signals.

Analytical studies for the application of various digital concepts to the instrumentation, voice, and television data systems became the focus of spacecraft communications. The main problem was still deciding which system of signal modulation was to be employed: analogue or digital. While the commonly held belief by engineers that digital was the most efficient means of signal transmission, when correctly employed an analogue system required substantially lower power consumption. However, as suggested in the earlier streamlining proposals, by removing non-essential picture information within the resource hungry signal, it was found that digital techniques could approach the analogue model in its competitiveness.

In view of the methods available for modulation of the TV signal, a considerable amount of investigation was made into the digital structure of data within the television blanking periods, that is, the point where no information of the image is being written onto the screen. After substantial research into the effects of reducing image quality, it was established that digital techniques were definitely feasible to transmit the low grade video. This however did not automatically rule analogue out, as the digital modulation required state-of-the-art equipment which could prove cumbersome and unreliable in a real mission.

Comparative evaluation of the analogue and digital systems was completed by 31 August 1962, and an analogue TV system was selected. The choice was made by the participating subcontractors using the following demonstrated and projected factors. Firstly, the digital system proved to require more development work to compete with an analogue system for transmission from lunar distances. The project subcontractors were unable to supply operating parameters in the time required. Secondly, as pointed out above, the complexity of the digital system was far greater than the proposed analogue system. A detailed study, entitled the Final Report for Manned Spacecraft Television completed and released nearly a full year later in July 1963, investigated the methods of digital television compression perhaps in a last ditch effort to ascertain whether the digital proposition was still feasible. The goal was to compile dependable estimates for equipment reliability, weight, size and power consumption for just such a digital system. A variety of compression techniques were evaluated in the study which included fifty-four different methods from the United States, England, the Netherlands and, surprisingly, the Soviet Union! Compression tests were carried out using selected scenes representing what it was expected the future lunar astronauts would encounter in their moon explorations. These analyses included spacecraft interior representation, exterior depictions of the spacecraft, exploration scenes (such as the lunar landscape) and standard television test patterns.

To compare the differences between analogue and digital television compression methods, similar design studies were performed on a variety of setups of digital television application and conventional analogue television processes. The end results looked very favourably on the future potential of digital compression. The standard results saw a 3:1 compression ratio which would directly result in the saving of transmission power and size, along with the successful miniaturization of the onboard television equipment. Indeed further recommendations strongly supported further investigation into the possibility of a 6:1 ratio.

Additionally, the proposal of onboard data storage was addressed. Compared against a rapid film processor, videotape recording emerged as the best alternative for data storage on the spacecraft. Based solely on discussions with industry representatives, the study envisioned a significant miniaturization of tape requirements by 1970. One company, Machtronics, went so far as to supply a possible recorder design as projected by them for use by the end of the 1960s! While the idea was feasible, it was hampered by weight and size, although, some fifteen years later, a similar system was successfully deployed in the CSM for the Apollo-Soyuz Test Project.

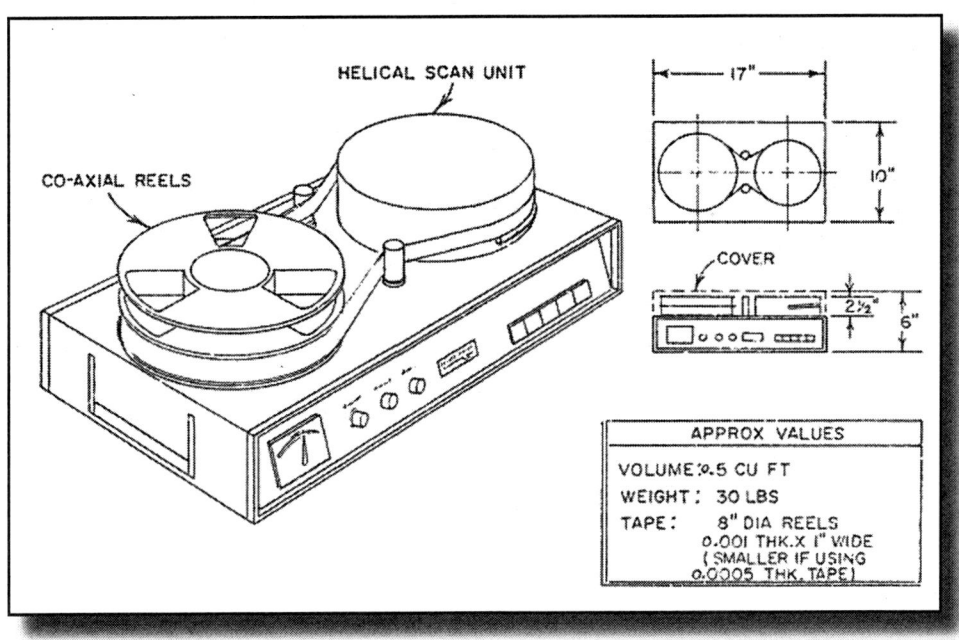

The proposed on board video recorder designed by Machtronics.

NASA had begun to review the necessary requirements and limitations of both analog and digital systems on the planned telecommunications equipment. Particular emphasis of these reviews was placed upon reliability, accessibility and cross-compatibility within the Command and Service Module. If the use of television on the Apollo missions was to occur with minimal departmental objections, then such procedures would have to pose absolutely no hindrance to important mission critical operations. Using the above-mentioned guidelines, combined with the tentative size, weight and location strategies, an acceptable video system to transmit pictures from the spacecraft to earth had been tentatively established by the Manned Spacecraft Center.

Once the theoretical issues had been cautiously prepared, NASA had a working model on which to base a practical system. At last the agency was in a position to delegate the contract for supplying the necessary television equipment to companies willing to take on the challenge of developing and furnishing the apparatus crucial to realization of live television from within the spacecraft and on the lunar surface. The Manned Spacecraft Center requested proposals for a television system from prospective outside contractors. However the initial response in reviewing the submitted proposals was rather disappointing and threatened to strangle the development NASA had hoped would occur rather quickly. On April 3, 1962 it reported that "Of the ten proposals received...no contractor met the full requirements of the specifications. Five of the contractors are completely unacceptable while the other five are only slightly varied from the specifications and are classified as acceptable."

Although it would be foolish to expect any phase of Apollo to have been fully flight-ready in the early months of 1962, the fact that none of the contractors could meet all the requirements was disheartening. It was becoming clear that the amount of research required for the portable TV camera system was far in excess of what NASA had originally envisioned, so much so that the ten proposals were again reviewed a little over one month later. Each contractor's submission was evaluated against NASA's criteria. No holds were barred in detailing which companies had the potential to successfully meet the needs of a lunar TV system, based not only on technical merit but also on experience related to the objectives at hand. General Electric, Fairchild Camera and Instrument Corporation, General Electrodynamics Corporation, IBM, and Raytheon presented their outlines which were subsequently deemed suitable by NASA.

Curiously, while the five successful contractor proposals were all technically acceptable, Raytheon surpassed all the other proposals in that it furnished a system of greater growth potential and state-of-the-art advancement capabilities, namely in the scan converter technology required to make the video signal conform to standard broadcast television equipment. Hallamore, RCA, Philco, Ball Brothers and ITT's proposals were determined to be technically unsuitable.

Furthermore, in additional discussions made on April 3, 1962 the Electrical Systems Board had noted that while acceptable, none of the tenders fully met the preconditions as laid out in the request for proposals. Even though the endorsed system of Raytheon was expanded to include General Electric as well, it was quite evident that considerably more investigation and testing would be necessary to fully develop a suitable television system for use in Apollo.

Nevertheless, even at this early stage of TV camera considerations, significant planning was given to the integration of TV cameras into mission operations. Unfortunately, the main drawback of the early studies was that ultimately no concrete mission profiles, television coverage timeframes, or completed equipment were available to provide any practical estimates to the study results. Other factors were also responsible in ultimately bringing the analogue system into favour. These issues were simplicity, a desirable signal-to-noise ratio, and reliability of operations.

While internal considerations regarding TV camera operations during

Apollo flights were being hotly debated within NASA, the public was anticipating a window via television into the heavens. Additionally public attitude was wholeheartedly behind the US Space Program especially if it meant beating the Russians after what had predominantly been Soviet victories occurring in rapid succession. Given this mindset of the United States competing against the Soviet Union, the momentum, along with financial backing, was at an unprecedented peak. Whether those within NASA liked it or not, television was expected to play an important role in realizing the goal of a manned lunar landing. The space race was now definitely on!

The NASA labs in which the TV equipment was tested for reliability and later flight qualification (NASA Photographs S68-48007 and S68-48008, next page).

CHAPTER 2: FAITH 7 SLOW SCAN TELEVISION

"This is going to practically be a flying camera."

L. Gordon Cooper, at his Mercury Faith 7 pre-flight press conference 1963

During the highly concerted effort to achieve the goal to fly men to the moon and return them safely to earth, NASA wasted no time in getting important experience in actual spaceflights. On May 5, 1961, a mere three weeks prior to Kennedy's bold and impassioned plea to Congress, Alan B. Shepard became the first American in space, flying in his spacecraft proudly named "Freedom 7". His 15-minute flight was part of the first manned spaceflight program of the United States called Project Mercury which had the ultimate goal to successfully place manned spacecraft into earth orbit.

The Mercury capsule was small and cramped, but served its function well and provided the necessary manoeuvrability for the pioneering spaceflights. In all, seven manned flights were successfully launched from Cape Canaveral in Florida and allowed the United States to gain crucial lost ground in the space-race against the Soviets. Each progressively longer flight of the program conducted an array of tests and experiments designed to give NASA vital data in the understanding of the effects of space on humans.

The last flight of the Mercury program was set to be the longest one in the series. Commanded by Gordon Cooper, the space capsule, dubbed "Faith 7" was to utilize a television camera during the flight. The main goal of the mission was to complete 22 orbits of the earth and to observe how an astronaut would cope with spending more than a full day in space. Due to the numerous orbits, it was an ideal opportunity to test the TV camera system with the ground stations which had been modified to receive the signal.

A similar television system had first been tested during the NASA Echo II prototype balloon tests in January and July of 1962. The passive communications satellite experiments consisted of a balloon satellite with a metal outer skin which acted as a passive reflector of microwave signals bounced off of it from one point on the Earth to another. The separation of the canister containing the balloon from the Thor-Agena rocket launch vehicle, the test sphere deployment and subsequent inflation were all highly desirable events which were observed in a suitable degree of resolution on the ground in real-time. The television camera observed these events unhindered in the vacuum of space as direct monitoring of the process by the naked eye was not possible.

The Lear Siegler supplied TV camera for the Echo tests operated at

30 frames-per-second and featured a resolution of 280 lines, which was more than adequate for good quality images. Slower scan rates had been considered, but the amount of motion information potentially lost at slower scan rates of the sphere deployment resulted in these not being used. The signal was transmitted via FM to conserve transmitter power, and the entire black and white TV system weighed in at 48 pounds and consumed 50 Watts of power! However as the tests were allotted a generous weight and power budget, there were no apparent problems for the mission.

The suggestions made for spacecraft television following the Echo II tests included recommendations that, with adequate planning and system streamlining the TV system weight could be reduced from 48 to 28 pounds. Such considerations had always been at the forefront of mission planning and were somewhat amplified by NASA's general reluctance for real-time television to be included on manned spaceflights.

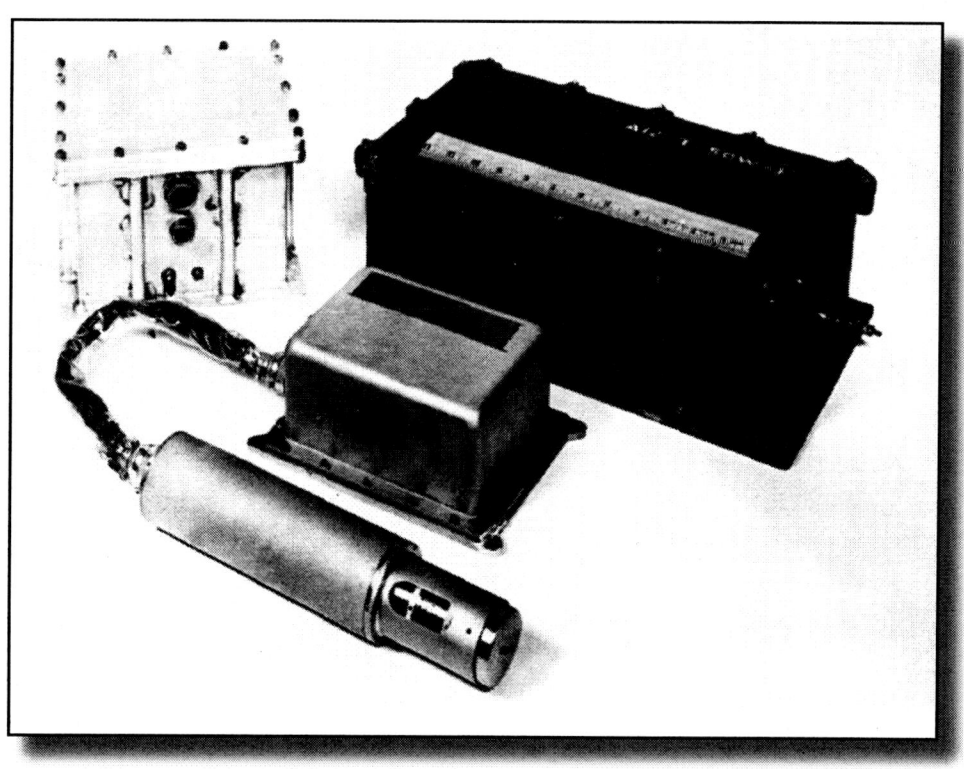

The Echo II Balloon Television System used to monitor the sphere deployment in space.

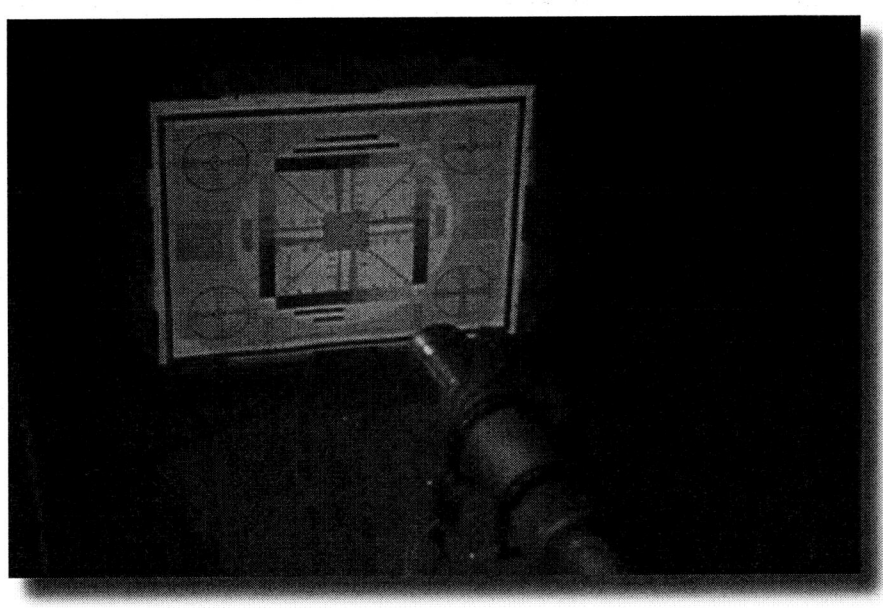

The TV Camera to be used on the Mercury Faith 7 flight undergoing resolution tests.

The MA-9 Flight plan from April 15, 1963 cited the following in regards to the TV camera carried aboard Faith 7, "A special TV camera will be carried on this flight to evaluate its operational value for monitoring the pilot's well being, for obtaining backup readings of the instrument panel indications, and for observing tests, experiments or external phenomena through the spacecraft window. This may determine the need and value of this technique for obtaining data in future manned spaceflights over methods now in use."

According to the flight plan a successful series of tests was crucial in shaping the direction NASA took in the use of a television camera on space flights. Cooper notes, "It was slow scan, but...we'd had a question whether or not it would even work from space or not. And so it did prove that it would." Curiously, the publicity factor was completely overlooked in the anticipated applications of the camera system. Cooper also related his own feelings regarding the desire of the public to feel some sort of connection with their space program, "I think most of us felt it was important to personalize the flight, that that's what really made the public get close behind and have a close feeling for those things they followed so closely." Unfortunately NASA management seemed to hold a different attitude, and cited that TV was only necessary in monitoring crew performance and spacecraft.

The Mercury camera which, like its Echo predecessor, was designed and built by Lear Siegler Inc. and arrived on March 19, 1963 at NASA for preparation for Cooper's upcoming Mercury flight. It weighed 10 pounds, consumed a rather hefty 56 Watts during operation and could be hand-held or bracket mounted inside the cabin just to the right of the 3-axis controller.

The TV camera mounted in the Mercury capsule.

The bracketing and necessary cabling added an additional payload of 7 pounds of weight, despite the system being remarkably compact compared to studio cameras in 1963. The camera had a scan rate of 1 frame every two seconds (0.5 frames per second) in order to deliver adequate picture quality while still remaining well within the size and weight constraints. Due to this non-standard scan rate, special equipment on the ground was needed to reconstitute the picture, and only the Mercury Control Center (MCC) in Houston, Coastal Sentry Ship (CSQ), or the Canary Islands (CYI) ground stations were equipped to receive the special TV signal. Additionally, in MCC the flight surgeon was able to view the TV in real time using a specially modified monitor which was capable of handling the two second scan rate. The use of scan conversion equipment was made and 5 additional TV screens were supplied with the scan converted images. CSQ and CYI were equipped to record the raw feed onto magnetic tape for later playback if it was so desired. Additionally, NASA opted to archive the downlink on 35mm film.

The operation of the camera was performed by the astronaut via a switch on the spacecraft control panel. The switch had three positions, "TV", "OFF" and "TM". By selecting the "TV" position Gordon Cooper would place the camera into operation mode and the TV system would be transmitting data. A test feature was also made which would place all the telemetry (or "T/M" as it was referred to during the Mercury missions) data using the telemetry low frequency transmitter onto the TV transmitter instead. This could be used as a backup in the unlikely event of failure on the main telemetry transmitter, and all ground stations used for the Mercury mission could receive this data.

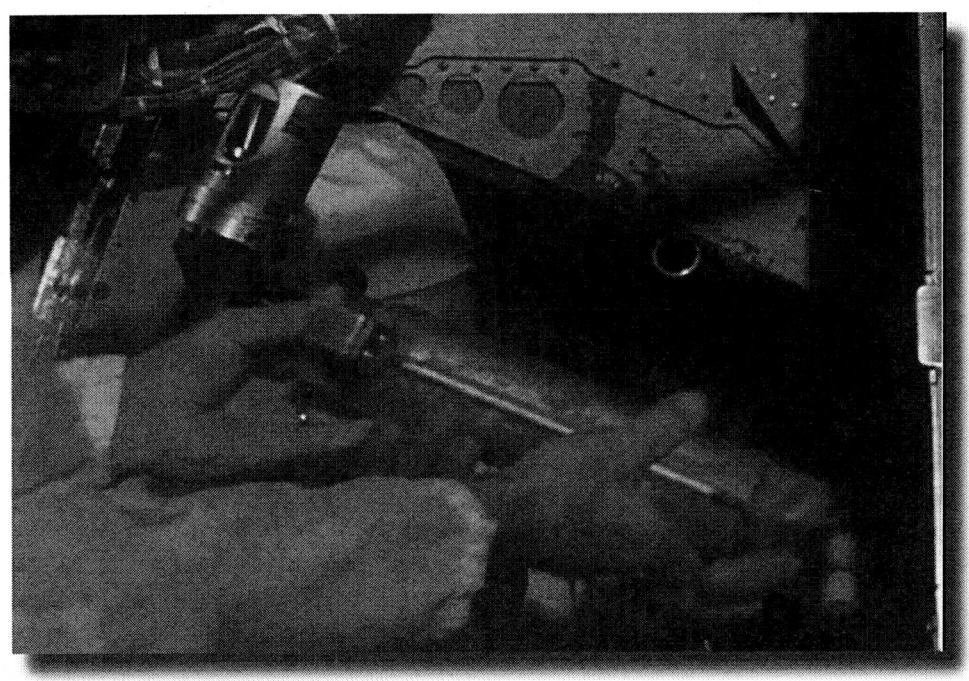

Faith 7 TV camera Close-up during mounting in the spacecraft cabin.

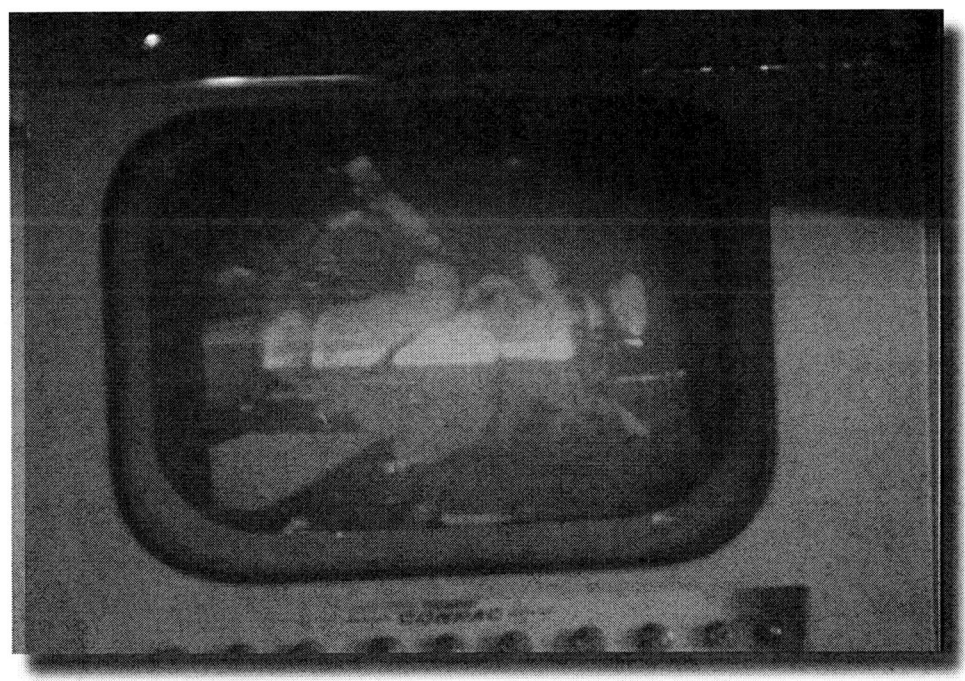

The slow scan Mercury TV image as seen on a monitor during camera tests.

A full test of the system was made while Faith 7 was flying over Hawaii and California during the 3rd orbit of the mission. To allow a wider degree of flexibility in what images could be shot by the camera, two lenses were provided. A 1" wide-angle lens provided a 26° field of view, while a 6" lens offered a 4.8° field of view. "This is going to practically be a flying camera," joked Cooper at a press conference on February 8, 1963, when he discussed the numerous photographic and TV equipment to be taken on the flight. Hopes were certainly high for television on the final mission of project Mercury.

Cooper's Faith 7 flight was launched on May 15, 1963 from Launch Complex 14 at Cape Canaveral, Florida. As planned, on board was the Lear Siegler television camera which, it was hoped, would meet all expectations during the flight. Throughout Cooper's 34 hours in space he made numerous broadcasts to earth using the camera in both its mounted and hand-held configuration. The following lists all the times he operated the camera during the flight:

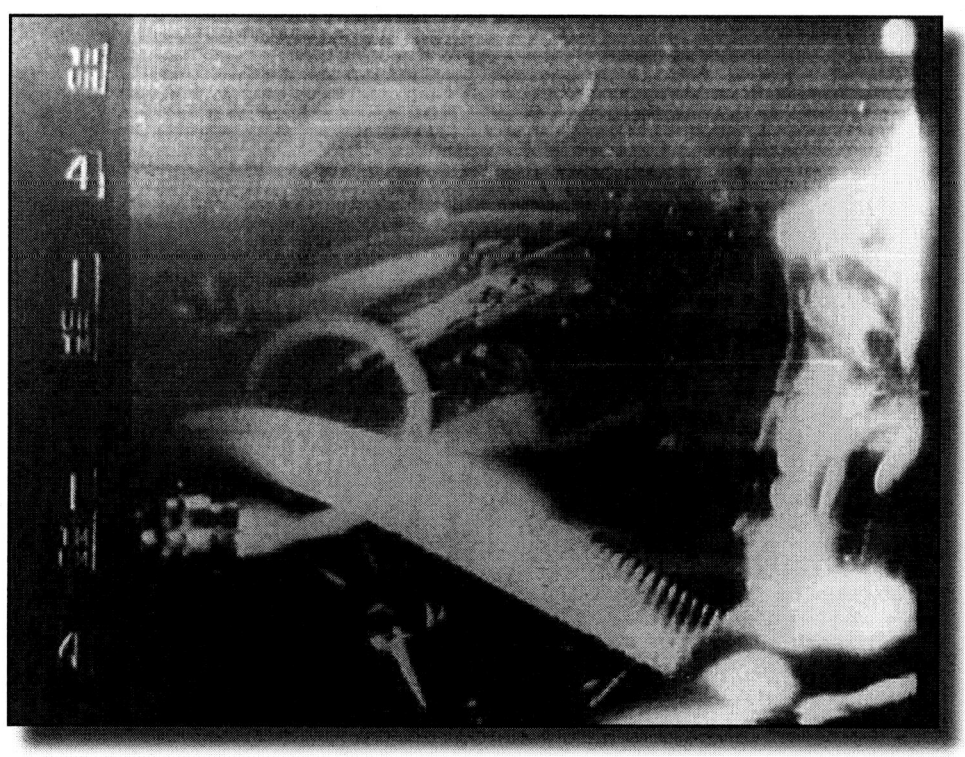

A 35mm print of a slow scan television frame (NASA S63-07856).

GET hh:mm	TV TEST AS PER FLIGHTPLAN	COMPLETED
0:07	Check TV OFF	YES
0:14	TV ON	YES
0:21	TV OFF	
1:34	TV ON	YES
1:40	TV OFF	
1:48	TV ON	YES
1:54	TV OFF	
3:07	TV ON	YES
3:14	TV OFF	
4:23	HAW T/M via TV check	YES
4:38	CAL T/M via TV OFF	
4:41	TV ON	YES
4:47	TV OFF	
6:15	TV ON	YES
6:19	TV OFF	
8:48	TV ON	YES
8:55	TV OFF	
10:22	TV ON	YES
10:28	TV OFF	
11:55	TV ON	YES
12:02	TV OFF	
13:29	TV ON	YES
13:36	TV OFF	
15:02	TV ON	NO
15:08	TV OFF	
17:38	TV ON	NO
17:43	TV OFF	
19:10	TV ON	NO
19:17	TV OFF	
22:03	TV ON	YES
22:09	TV OFF	
22:17	TV ON	YES
22:24	TV OFF	
23:36	TV ON	YES
23:43	TV OFF	
23:51	TV ON	YES
23:57	TV OFF	
25:10:00	TV ON	YES
25:16:00	TV OFF	
25:25:00	TV ON	YES
25:29:00	TV OFF	
26:43:00	TV ON	NO
26:49:00	TV OFF	
28:16:00	TV ON	YES
28:23:00	TV OFF	
29:51:00	TV ON	NO
29:54:00	TV OFF	
30:53:00	TV ON	NO
30:54:00	TV OFF	
33:57:00	TV ON	NO
34:03:00	TV OFF	

TV Transmissions both planned and completed during Gordon Cooper's faith 7 Mercury flight.

Cooper related his use of the TV camera by saying, "The TV System appeared to work satisfactory at first. Then I don't know what happened, whether it gave up completely, or just how they fixed it toward the end. They didn't appear to be getting much at the end." When the tests were going successfully, the images captured by the camera during the tests were Cooper's portrait while sitting in the cockpit, and views out of the spacecraft window while flying over certain interesting land features. During his flight debriefing he noted, "I used it for pictures with it in the bracket and out the window. It took a great, tremendous physical conditioning process to get the camera out of the bracket and get the other lens on it prior to photographing out the window. I got one period of pictures out the window over Florida that was exceptionally good. In one, it was ideal weather, ideal view, and everything, however, they weren't getting anything at all."

The not-so-spectacular performance of television during the flight of Faith 7 did little to change the anti-TV opinion which seemed to sweep through NASA management. Even astronaut Gordon Cooper, who was well aware of the need to bring the public intimately close to the U.S. space program, seemed to offer little praise for the TV camera he had tested during his flight, "The attachments were easy enough to handle, but the TV camera was difficult to use where it was located and really quite burdensome. The cable on the camera, incidentally, worked out well. The way the cable was rigged, it was well out of the way of the control stick. But you kept knocking the control stick all over the place every time you reached over to get the TV camera or to put it back in place. So I always powered everything down before I ever moved the TV camera." He further clarified the procedure of the TV signal acquisition and setup by stating, "It didn't seem to be difficult at all. However, I didn't seem to get any information as to whether they were really getting anything down on the ground or not, which I suspect they weren't half of the time. Numerous times they told me they weren't."

Certainly, following the trouble experienced in getting an acceptable picture from Faith 7, the opinion of many within NASA was that the camera had most definitely not performed as hoped. This certainly played a decisive factor in the exclusion of TV cameras from the upcoming Gemini project, which would test many of the new spacecraft manoeuvres ultimately planned for the Apollo lunar voyages. Cooper's own testimony reinforced the direction NASA took in disregarding live TV from the orbital missions of future manned spaceflight, "I could see no real advantages to the pilot in having it onboard. If the thing had worked correctly, it would have been nice to have been able to shoot out the window at some of these fantastic views you see around Florida and around CSQ to show the ground what you're seeing. That would have been the biggest advantage of it. I gather, based on what limited information I have now, that we got one or two successful pictures of this type, but the camera is hard to use where it was located. It's very difficult to get out of the bracket. You have to plan ahead quite a ways to start getting it out of the bracket and there's no place you can put it if you want to use anything else while it is out."

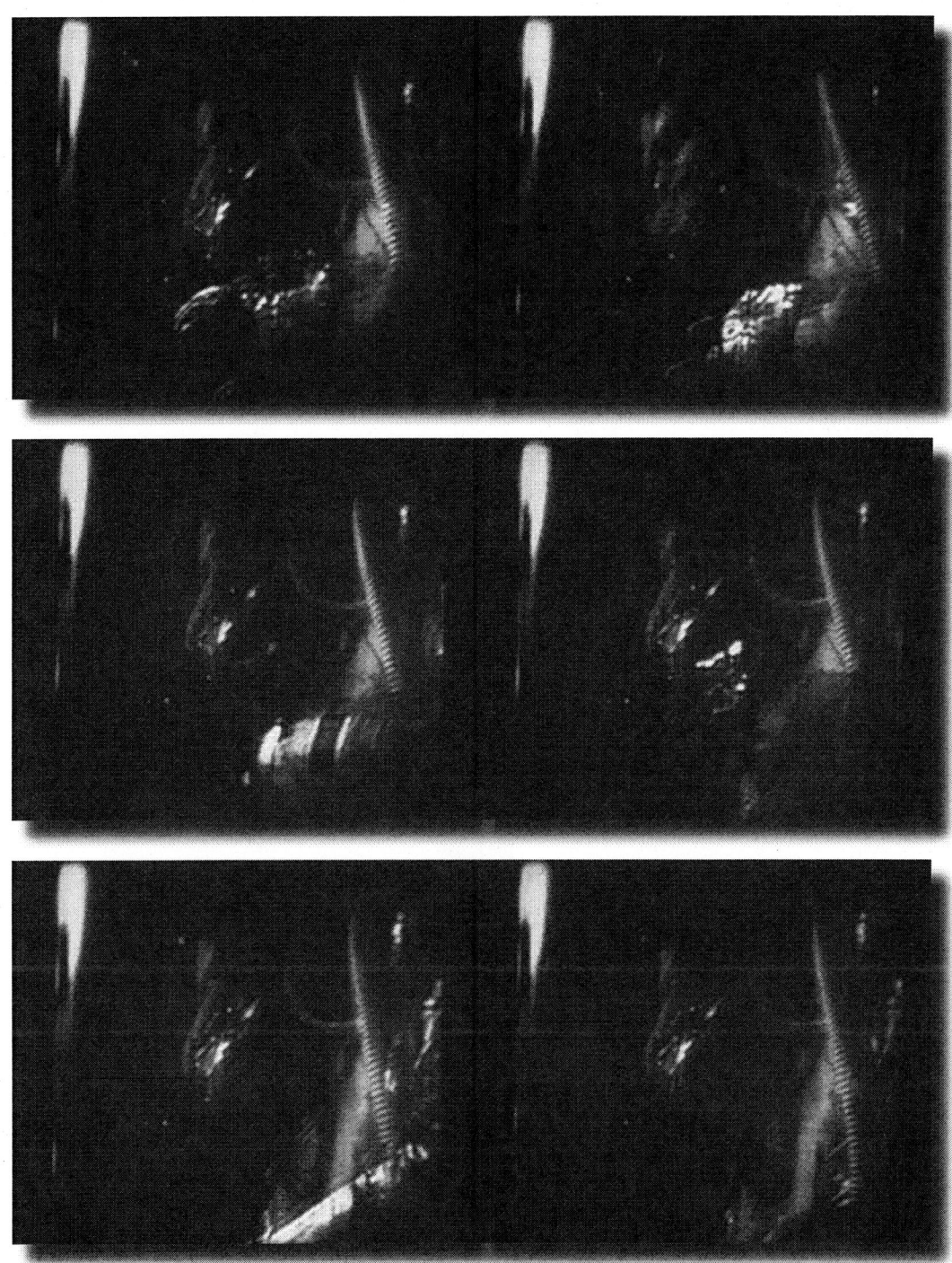

The only known surviving frames of Gordon Cooper during the TV camera test on Faith 7, made while flying over Florida (From the NASA film "The Flight of Faith 7").

"You could fly the spacecraft with the TV camera out. I flew it on manual proportional or fly-by-wire several times to establish a very slight change in attitude while I was using the camera. No problem there, but you can't use another camera, for instance, while you're using the TV camera. There's nowhere to put the big, bulky rascal, and you've got to take that 6 inch lens off prior to storing it back in the bracket. Which is a mess. You have to take it off, put it somewhere, then get the 1 inch lens, put it back on the camera, then stow the camera, and then stow the 6 inch lens."

The need for a simplified easy-to-use and compact camera had become much more important following the lacklustre of Faith 7's in flight tests. The future of television from space depended on it. Not surprisingly, NASA turned to the electronics company RCA to seek resolution to the spacecraft television system dilemma. If the television cameras to be used by astronauts in a spacecraft journeying to the moon, or from the lunar surface itself could not meet the system limitations outlined by the earlier television studies, then there would absolutely be no television at all on the Apollo spaceflights.

CHAPTER 3: RCA AND THE BLOCK I CAMERA

"[Television] will also afford a vicarious participation by the television audience in the most exciting and significant exploration of modern times and perhaps of all time."

John Lowrance in the February 1965 issue of the SMPTE Journal.

Despite the unfavourable review of the of the Faith 7 TV tests, the development of the lunar mission television system advanced at a breakneck pace. On 21 December, 1961 Collins Radio was awarded the contract for the telecommunications system for Apollo. Collins Radio, in turn for sake of simplicity handed over administration of the TV camera to North American Aviation (NAA), the contractor chosen to build the Command Module. Nearly buckling under the strain to supply all aspects of the CM within the allocated deadline, NAA contracted RCA Princeton Laboratories to supply the TV Camera. During the design of the CM TV system by RCA, NASA formed a group to oversee the administration of the television system which, enticed by the huge advances in top secret military applications of low light camera technology by Westinghouse, subsequently approached that company in late 1964 to design the lunar surface TV camera. Consequently two rivals worked in parallel for two distinct types of cameras. One of them was made for the Command Module, the other one for the Lunar Module and lunar surface.

In late 1961 NAA had been awarded the contract to build the Command Module for the Apollo Program. Preliminary launches were to use a Block I spacecraft to carry a three-man crew into space. The Block I CM was in effect a prototype design which did not include any mechanism to physically dock with the LM and so the design was never envisioned for anything other than low earth-orbit missions. Later actual lunar-landing missions would use the updated Block II design which allowed the LM to be coupled with it and allow crew movement between the two craft while docked.

Early in spacecraft planning the task of developing the analogue CM TV system was transferred from the communications contractor Collins Radio to NAA, which then awarded a subcontract to RCA. The cameras were classed as Contractor Furnished Equipment, meaning that RCA was solely accountable for its successful design and delivery. Not surprisingly, the design needed to be consistent with the recommendations made under the TV system studies conducted earlier by NASA.

When NAA was awarded the CM contract, they had originally estimated that under their management, the RCA TV camera would be delivered sometime at the start of 1963. Efforts were underway to assemble a camera capable of meeting the requirements NASA had established for the TV signal the year before. Throughout their Apollo program involvement, NAA published monthly progress reports detailing all aspects of the spacecraft development including equipment yet to be completed. In a report covering 16 February to 15 March, 1963, the delivery date for the TV Camera was expected to be met only one month later! Certainly the pressure was mounting in the pursuit of meeting such short deadlines. Curiously, after this date had passed without any resolution, the scheduled time was pushed back to May of the same year. It quickly became apparent that NAA would not be able to simultaneously design the spacecraft and oversee the production of the RCA TV camera within the projected timeframe.

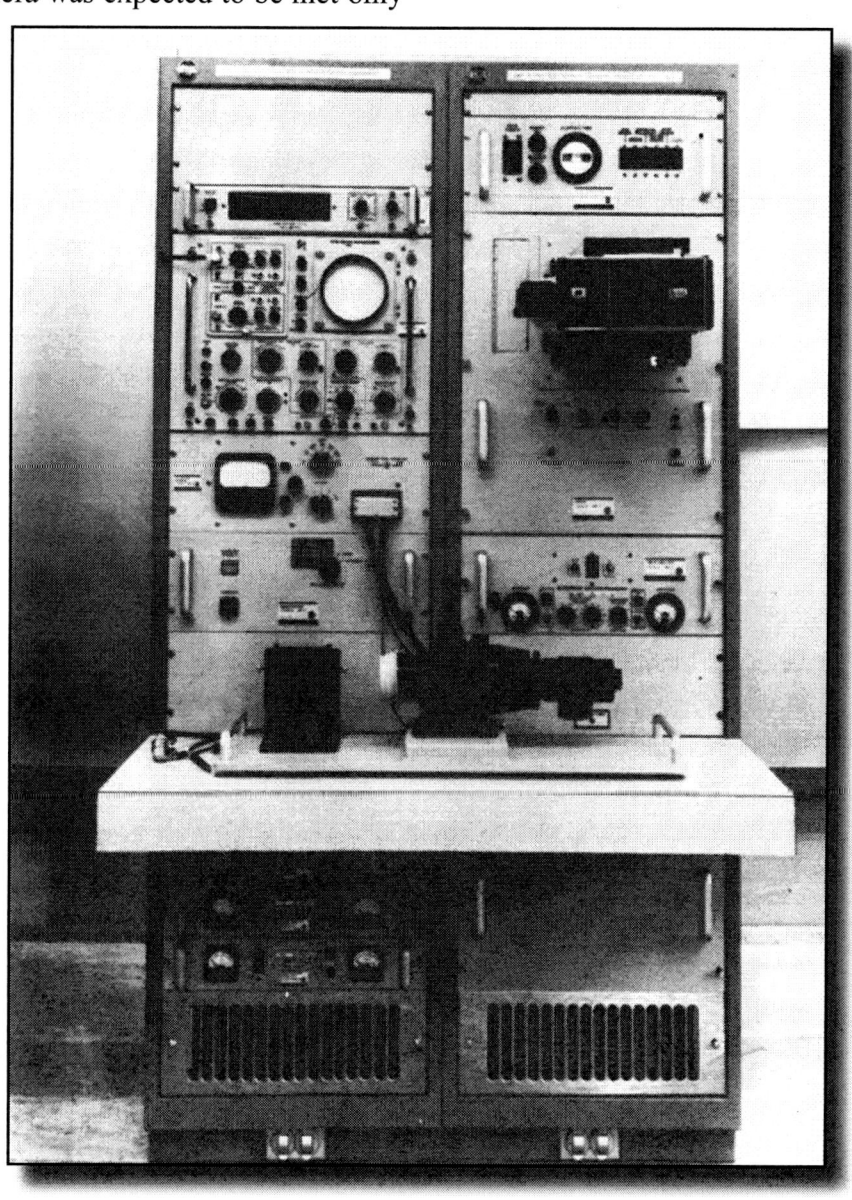

Television Subsystem bench maintenance equipment.

The July 1, 1963 NAA progress report in addition to finally announcing RCA Princeton Laboratories as the contractor for the TV cameras, mentioned that, "As a result of a meeting with NASA, Grumman, and RCA, the basic procurement specification for the command module TV camera has been revised to make it compatible with the Lunar Excursion Module [LM] TV camera. The required changes are an additional oscillator in the camera to provide sync instead of using an output of the central timing equipment, conversion from 115 volts ac, 3-phase, 400 cycles per second operating input voltage to 28 volts dc, alteration of the picture sync format, and reduction of the lines per frame from 400 to 320." Reduction of the electrical systems burden, that the TV camera would place on the spacecraft resources, necessitated the ruthless power-saving course of action. Furthermore, the signal changes as discussed between NASA and RCA would help facilitate these cutbacks by lowering the transmitter power requirements needed to send the TV signal back to earth. The fine balance between necessary minimal resource drain and an acceptable moving picture was continually being debated. As such, delays in providing the completed TV camera became quite evident in 1963.

One week following the July progress report, and after further discussions between NAA and NASA, LM Spacecraft Project Officer, J.L. Decker issued a memorandum in which he reviewed the current status of the TV camera to be used inside the CSM. He bluntly pointed out that given the enormously tight schedule placed upon NAA to deliver the Command Module, there was significant effort being wasted by them in trying to additionally supply a high-grade television camera. The argument was made that because of this, necessary changes required of the CSM camera could not be implemented without:

- Severely delaying its expected delivery

- Rigorously limiting the LM camera system capability

- Completely redesigning the entire television system.

His recommendation was that NAA abandon its plans for a CSM camera, and that the LM camera should become a general purpose TV camera for the two spacecraft. Given that RCA was at that time still the supplier of both cameras, the expenditure accrued to that point would not have been wasted, and in fact the future cost outlay for the television equipment would have been effectively halved as only one camera would be required in addition to a weight reduction of 13.2 pounds. Decker's closing recommendation was that the TV Camera portion of the NAA contract be cancelled and fall solely under the management of RCA.

While it appears no authoritative moves were made to cancel the NAA TV contract in the interim period, NASA did eventually form a group which took over the procurement of the television cameras from NAA. This group however, concentrated on approaching Westinghouse to produce a monochrome camera which could withstand the extreme lunar surface conditions. It is clear that following the internal discussions, RCA still concentrated on a single camera design for the entire Apollo TV system. It appeared that RCA was unaware that negotiations between NASA and Westinghouse were being conducted, and thus continued to work on a camera for both spacecraft. Within the same month further proposals were announced which would require the lunar module camera to, "…incorporate the capability of taking and transmitting TV pictures directly to earth during descent, ascent and lunar operations." The result was a new set of timetables for equipment testing in which reliability objectives were expected to be met and completed. These were subsequently to be made quite flexible given the anticipated new data which would inevitably present itself once proper benchmark tests had begun.

The Apollo Technical Manual with the subheading "Reliability" from October 15, 1963 stipulated that "The Apollo television camera has been designated nonessential to mission success or crew safety. However, since periodic monitoring of the crew throughout the flight could have ramifications in these reliability areas, a reliability requirement of 0.999 has been imposed on this equipment for a minimum of 40 flight operating hours. The selected supplier of the camera has predicted a reliability closely approximating this requirement. Several non-standard components must be evaluated and the design finalized before the actual reliability can be assessed…"

Television Communications workbench.

Block II CSM production had started with a concentration of efforts on the concept of lunar landing missions, thereby significantly changing spacecraft design and the need for a camera able to operate through all phases of such a mission. The Block II spacecraft would be equipped with a docking component allowing it to mate with the LM and allow the crew to move from one spacecraft to the other. The LM could then disconnect and journey to the lunar surface.

As documented in the numerous television system studies for NASA, the RCA Apollo TV camera was made with a series of design considerations different to those of a standard studio television camera. Simplification and size reduction, along with the reduced signal bandwidth were the prime limiting factors which had guided the RCA camera design. The main compliance issue of the Apollo TV signal was that, despite originating from the spacecraft with a lower frame-rate than those originating from studio-type cameras, it was required to be compatible with the NTSC television standard. Ground-based scan converters would adapt the TV image to a fully compliant NTSC signal and the converted signal would then be handled by regular broadcast equipment at Houston and subsequently sent to the major television broadcasters. RCA had developed a suitable camera for the purposes of providing real-time TV images from space, while operating well within the specifications drafted by NASA.

Technicians in the RCA Space Center in Princeton, New Jersey, assemble portable TV cameras smaller than a carton of cigarettes.

Near the end of 1964, RCA engineering team manager, John Lowrance, submitted an article to the Society of Motion Picture and Television Engineers in which he outlined the RCA TV cameras designed for the upcoming Apollo spaceflights, specifically noting the ways in which the cameras differed from studio-type cameras. They were lightweight, black-and-white and produced a picture resolution of 320 lines of information. Additionally, they operated at 10 frames per second, which necessitated the use of scan converters in order to make the signal compatible with millions of television sets in the United States - the prime target audience of the lunar missions. Lowrance further reiterated the three mission requirements for the TV camera.

These factors were predetermined by previous arduous studies which sought to reduce system resource reliance and spacecraft power, and from studies which minimized intrusion into the astronauts' work schedule:

- The cameras were to withstand the shock induced during launch while mounted in a position which could monitor the crew

- After launch the camera would be relocated behind the couch in a position which could monitor the centre aisle and work area of the spacecraft.

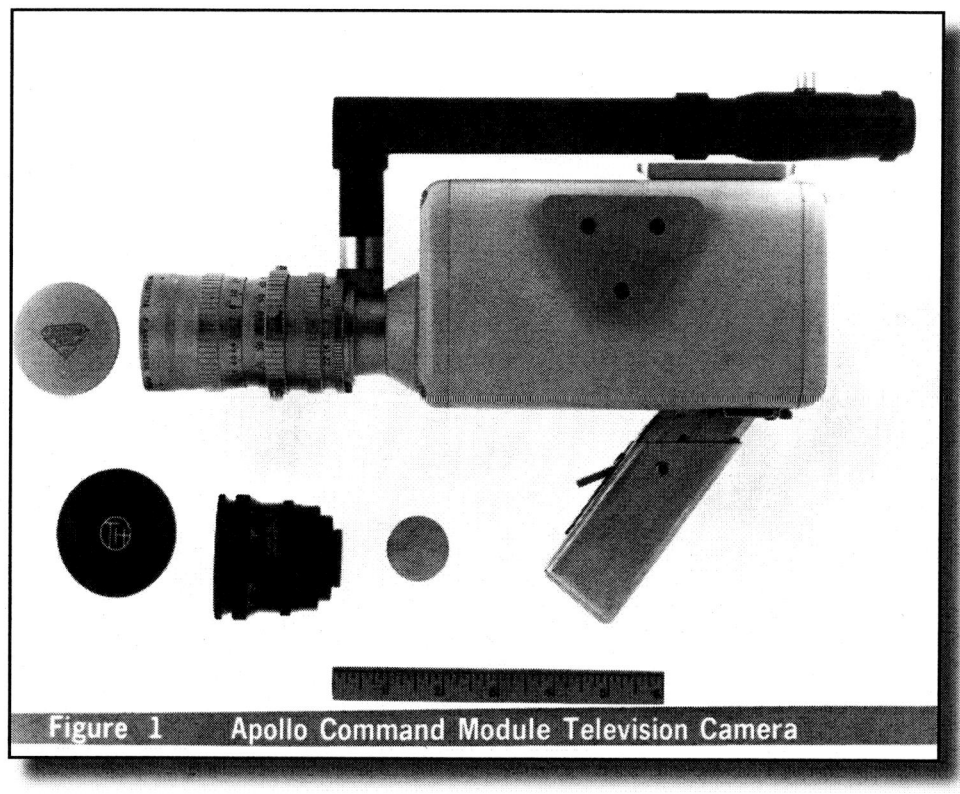

Figure 1   Apollo Command Module Television Camera

(NASA Photo 64-H-2684)

- Handheld operation could easily be performed to allow shooting through the spacecraft window, or at anything requiring closer inspection during a telecast.

An important physical design consideration of the TV camera had been weight, while performance and reliability were also given high priority. Because of these stringent requirements, a modified RCA-8134 Vidicon tube was the only option considered by RCA for the image sensor inside the camera. The Vidicon was developed by RCA as a compact and simple pickup tube for small, lightweight and transistorized TV cameras. The pickup tube in TV cameras is the section which converts the light coming through the camera lens into an electrical impulse which can be then transmitted via radio waves. Within the television industry the 8134 tube was already recognized as being at the forefront of miniaturized component design and was well suited for pickup in low-light situations. The image tube inside the RCA camera used an additional Silicon Intensifier Target (SIT) which greatly increased the camera's already superb sensitivity to light. The modification had the photoelectrons generated at the image sensor focused onto an additional silicon diode target which reacted more effectively to the electrons, thus making it more responsive to low light.

The camera power requirements were a further consideration guiding its design. Alongside more effective power consumption considerations, integrated circuits were used wherever possible to reduce the size of the camera. In certain camera functions, the unavoidable use of hybrid circuitry was employed as it offered the best solution for practical design of the unit. A total of 45 circuits were built which controlled the TV camera's operations. The end result was a camera which was hoped could also be operated by a fully suited astronaut with very little effort.

An inside view of the RCA Block I TV camera.

The perceived strain on the spacecraft crew had long been a sore-point in television integration on Apollo spacecraft, despite RCA's efforts in simplifying the camera. However, the near disaster encountered on Gemini 8, whereby Neil Armstrong and Dave Scott nearly lost full control of their capsule apparently prompted Robert C. Seamans, Deputy Administrator to carefully reconsider the anti-television feelings within NASA.

On March 30, 1966 he issued a memo in which serious non-public interest considerations were given to the idea of real-time television during Apollo spaceflights. He writes, "As a result of the Gemini 8 experience, I have been reflecting on the network coverage for Apollo." He then states that, "The television equipment would not be solely used for news purposes, but undoubtedly all manner of demands will be placed upon us for continuous live coverage of the lunar expedition." His concern for the safety of the crew and for the ability of Mission Control to provide instantaneous support should the astronauts encounter difficulties during the flight were high enough for him to request meetings to resolve the question of feasibility of TV during all phases of the journey (excluding the period when the spacecraft was behind the moon and out of contact with the earth). He requested ideas on:

- The capability of the present equipment to fulfil this requirement

- Plans for network coverage

- Probable modifications required on existing equipment in order to provide better means of TV coverage during the journey.

This memo stirred up activity once again for research into the use of television during lunar expeditions. A July program review specifically mentioned the cost and suitability of television coverage for both earth-orbital and lunar Apollo voyages, which in turn prompted further investigation into methods available to implement such coverage at minimum hindrance to mission objectives. A response by George Mueller issued on November 9, 1966 listed the considerations and projected the estimated time such measures would be fully operational for use and/or testing on an actual flight.

Mueller noted that by incorporating the coverage provided by both Madrid and Goldstone tracking stations between 62 to 91% of anticipated TV time during the mission would be made available, with an initial cost of $500,000 plus an annual running cost of $1,200,000 per year based on four week-long missions which would have required scan conversion. He also argued against using Canberra, highlighting that the start-up cost would be six times greater, with running costs effectively doubling the annual operation costs for little benefit in extended coverage. Mueller was also quite firm that the entire system be fully tested during an actual mission at least once prior to a lunar mission. While the implementation of receiving video to be scan converted was not yet instigated, he was confident that limited TV broadcasts from the capsule would be available during the flight of AS-204 (Apollo 1).

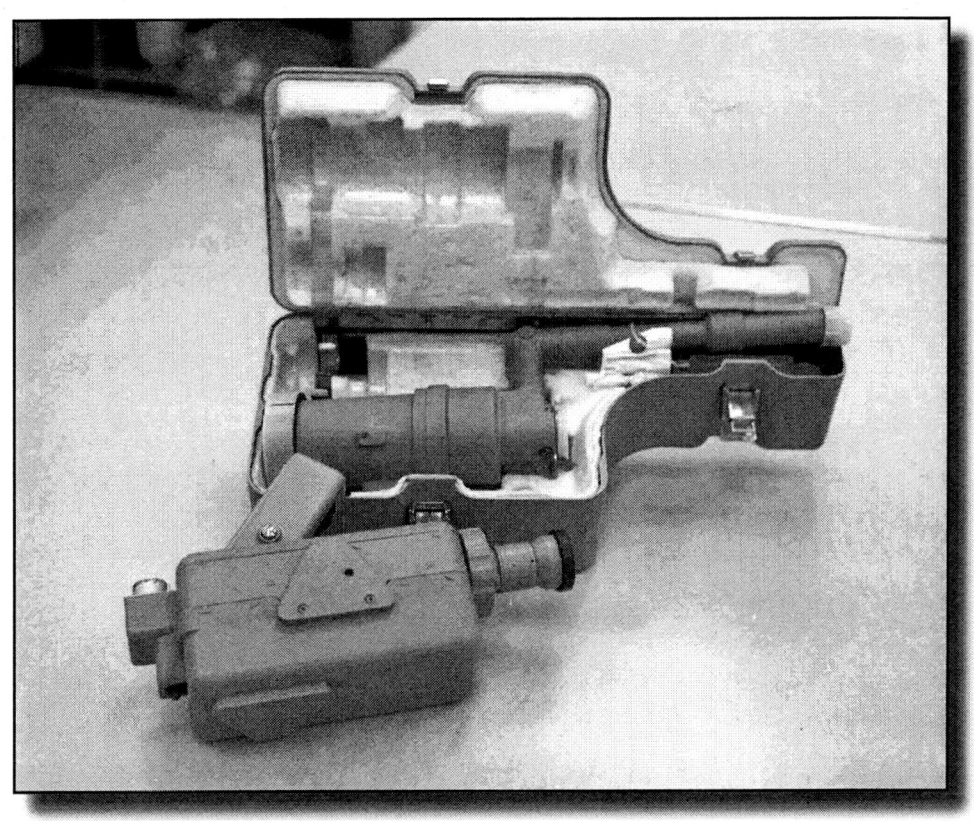

**The Block I RCA TV Camera and its protective storage case.
(NASA Photo S67-15357)**

Christopher Craft added to the discussion on TV use for AS-204 by clarifying exactly what NASA should expect from the broadcasts anticipated during the first manned Apollo mission. He cited that it would be roughly 50% of the quality seen on commercial networks, with some jerkiness of motion due to the frame rate of 10 frames-per-second. In a handwritten addition to his report, he also emphatically stated that there were, "… no provisions to display this info [TV] in real time at the MCC-H [Mission Control Center Houston] nor any other MSFN [Manned Space Flight Network] site. All we presently plan is the recording of the signal (if used) for post flight analysis." Kraft in summarizing did, however, suggest that if George Low had any recommendations regarding the need for real time TV coverage he would certainly be open to any recommendations.

While the mood within NASA suggested that live TV was more a nuisance than a benefit, it was becoming apparent that there would be transmissions made even if they were reluctantly supported by management.

The RCA Block I television camera and a test of the camera through the RCA slow scan TV signal converter. (RCA Astro Electronics Division Photo 68-12-08C)

CHAPTER 4: AS-204 - APOLLO 1 TV PLANS.

How are we going to get to the moon if we can't talk between two or three buildings?

Gus Grissom, moments before the Apollo 1 fire, January 27, 1967.

The ill-fated mission of Apollo 1, or AS-204 as it was originally known, was planned as the first manned flight of the Apollo Spacecraft, scheduled to fly some time on or around 21 February, 1967. Commanded by veteran astronaut Gus Grissom and accompanied by crewmates Ed White, who had completed America's first spacewalk and Roger Chaffee, a space "rookie" on his first flight, the spaceflight was to be an all-out test of the Apollo Command Module in earth orbit. Like most things on the intended flight, the use of television was expected to fully test the system for reliability while also affording audiences around the world a brief glimpse into life on board an orbiting spacecraft. A film documented simulator run with the crew and Mission Control conducted in 1966, supplied visual confirmation of television camera operation during varied phases of the planned flight. Both the preliminary and final flight plans detail activities which included transmission of live pictures from within the spacecraft. What follows is every known sequence of events which would have taken place on the flight according to the available flight-plan from 9 December, 1966 and several NASA issued memos. The Apollo Operations Handbook section on the television subsystem, from 12 November, 1966 highlighted the following aspects for use of the TV hardware.

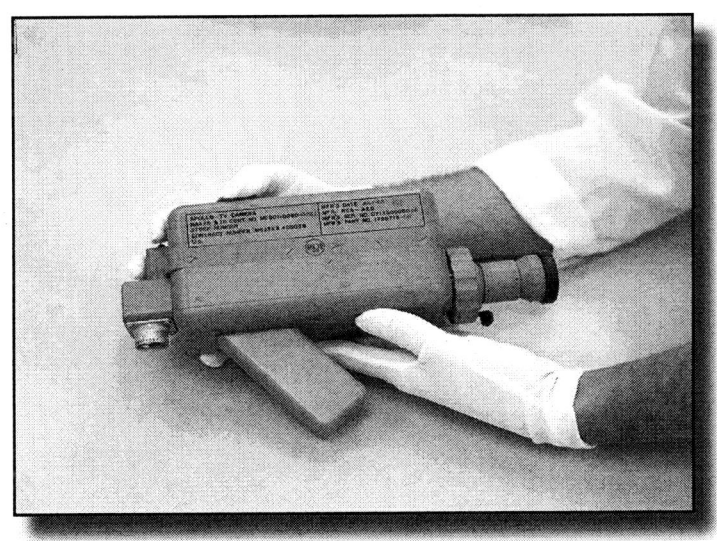

The Block I RCA TV camera planned for use on the flight of Apollo 1 (NASA photo S67-15358).

The television equipment which would have been used on Apollo 1 consisted of the Block I configured RCA black-and-white portable TV camera either mounted in one of four locations inside the spacecraft or hand-held via use of a variable-length pistol grip. Its prime function was to acquire real-time video information which would be transmitted from the Command Module to the Manned Spaceflight Network. The camera was able to be mounted in the following positions:

- Below the main display console which would have provided a frontal view of the crew while in their couches.

- Behind the top of the centre couch which would have allowed observation of crew performance in the middle of the spacecraft.

- In the right-hand equipment bay where this would have provided a view of the astronauts working in that area.

- Mounted on a special bracket allowing it to look out of the right-handing docking window.

- Hand-held operation would have been supplemented by a stretch cable allowing views of any object inside the spacecraft (and outside from any of the spacecraft windows).

When in its first position below the main console, the camera would have been connected directly to the power cable. Additional power outlets were available for connection via the stretch cable in whichever mounting configuration the astronauts needed to use.

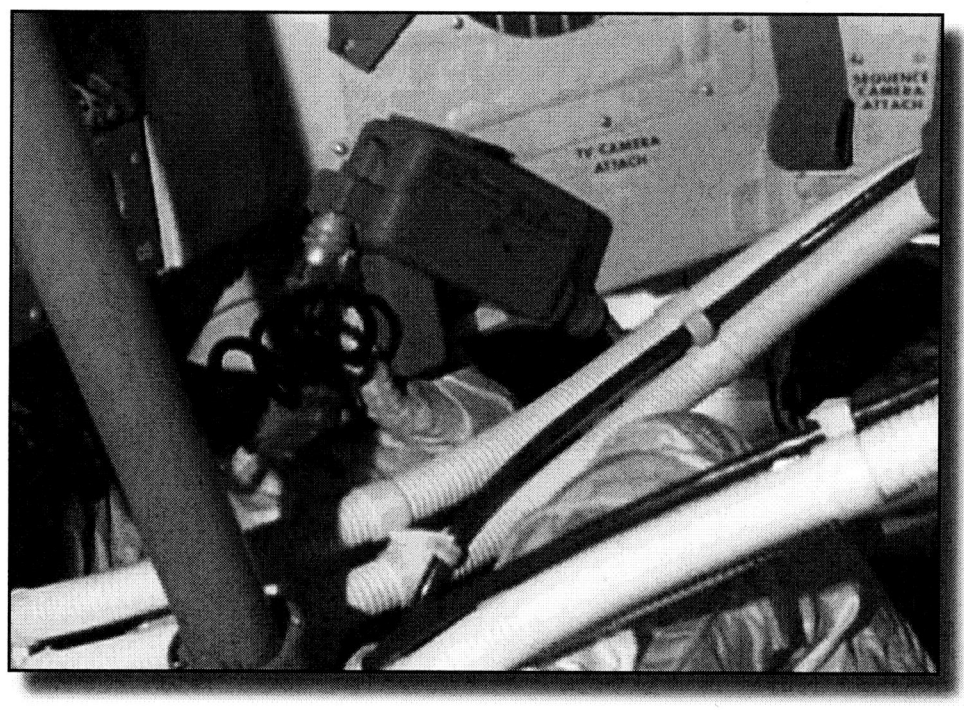

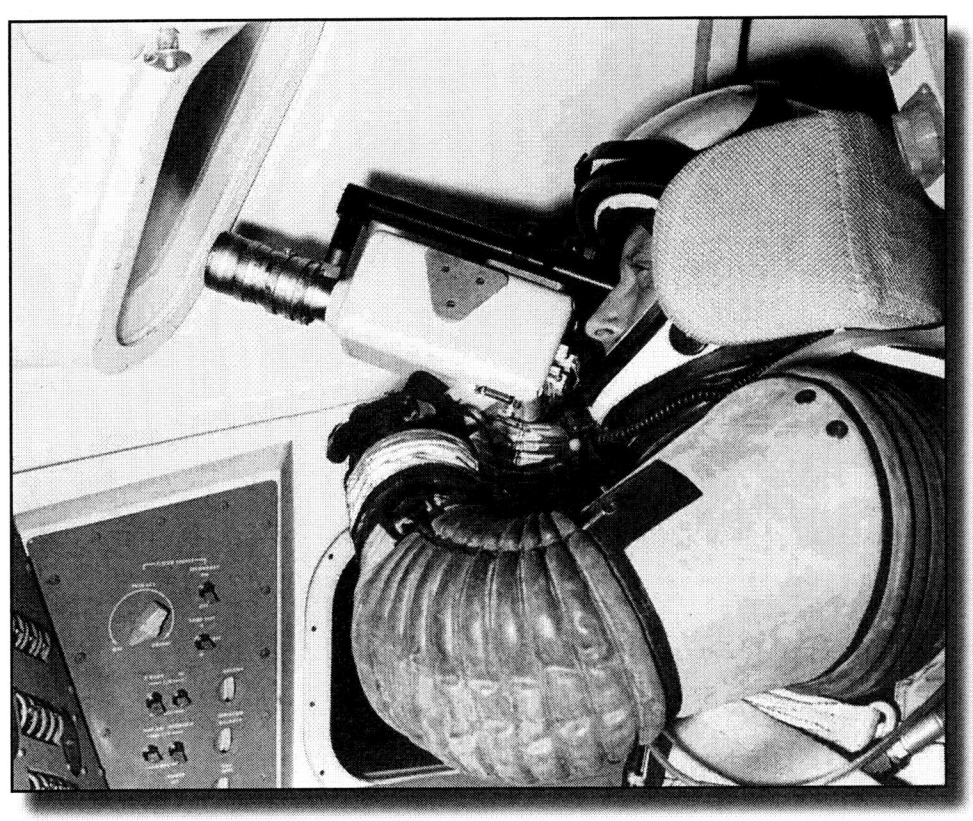

Anticipated locations (left and above) of the RCA Block I TV camera (from a 1966 training run and NASA Photograph S64-18109).

The pistol-grip facilitated portable use and was adjustable to comfortably fit the hands of the crew members. For telecasts inside the spacecraft, the camera was to be fitted with a fixed focus, wide-angle lens allowing sufficient views of astronaut activities despite the cramped space. An interchangeable zoom lens was provided for expected views of the earth or the moon from the spacecraft window. A through-the-lens type attachment was fitted which would allow the operator to see the changes in the image during the necessary lens adjustments. To supplement the varied lighting conditions which would have been encountered, especially between the interior and exterior views, the internal camera electronics featured an automatic gain control to tolerate the differences in light intensities.

The TV camera was activated by a slide switch in the pistol-grip handle, which applied 28 volts DC to the camera power supply when set to on. This powered the entire camera and resulted in a video output signal that was to be fed to the pre-modulation processor. Here, the video signal was to be combined with the telemetry data and voice signal, and subsequently transmitted to earth via the USB transmitter. The transmitter would need to be operating in a selected TV mode to allow this transmission to occur which the world would have eagerly anticipated once the mission had begun.

The RCA Block I TV camera held by an engineer in front of the Block I Command Module. (NASA Photo 68H-537)

Had the spacecraft successfully launched, 1 hour and 19 minutes into the flight or Ground Elapsed Time (GET), the Apollo 1 crew were to have unpacked and setup the TV camera, along with unpacking and attaching the required lenses, bore sight, spacecraft mountings and power cable. After camera preparation, no further activity would have been required for at least the two hours following this procedure until acquisition of signal over Guam, during the placement of the S-IVB rocket stage into retro attitude. This stage of the Saturn rocket would eventually house the lunar module on later flights to the moon (except Apollo 8), and also on that of Apollo 9, which fully tested the LM in flight. On the flight of Apollo 1 the stage would simply be observed from the spacecraft window. Once the camera would have switched on at 3 hours and 9 minutes GET, Mission Control was scheduled to advise the astronauts which type of lens would best suit the observation of this essential procedure. Possibly occurring during actual televised activities, the camera would have been adjusted to provide the best possible picture quality for reception on the ground.

The Apollo 1 crew during training. Roger Chafee is operating the TV camera.

The S-IVB would have then been commanded to start a 20 second Liquid Oxygen vent and a 36 second Liquid Hydrogen vent, all of which would have been captured via the RCA camera and sent to Houston. The flight plan stipulated that this observation would have alternated between TV coverage and standard still photography, because the Hassleblad images would have yielded much better detail for post-flight analysis of the procedure. The brief TV coverage would have lasted only 7 minutes before the spacecraft was out of range of the Merritt Island tracking facility. Following the use of the TV camera, it would have been stored at the Lower Equipment Bay with a wide angle lens to monitor the crew had Houston requested it.

Given the low earth orbit the spacecraft would have been in, all of the planned television transmissions from Apollo 1 would have been of short duration. TV images would only be obtainable as the spacecraft flew over one of the many tracking stations on earth. In some cases overlap of signal may have been possible, but it seems very likely that any broadcast would be restricted to a little over 5 minutes each time.

A full day on earth would have passed until the TV camera would have been removed from its position at the Lower Equipment Bay and mounted to the hatch. The crew would have then conducted humidity experiments via an aerosol particle analyser. The device measured content of microscopic bits of solid or liquid matter in the spaceship cabin for its possible effects on the well-being of the astronaut crew or the reliability of the electronic equipment. The engine burn, less than a minute after camera power-up, would most likely have also been observed during the television downlink.

At 69:00 GET the crew would have been expected to prepare the TV once again for operation. The activity covered would have more than likely been the essential change of the Lithium Hydroxide canister. These canisters allowed purification of the breathable oxygen for the crew and regular changing of them crucial for the crew's ongoing survival. As nearly demonstrated on the crippled flight of Apollo 13, failure to regularly change them could have quickly resulted in crew asphyxiation as the carbon dioxide expelled from the astronauts' lungs would have accumulated in the spacecraft cabin. Following the canister exchange, the astronauts may have demonstrated for viewers at home some more of the routine, yet crucial maintenance, which occurred during spaceflight. Additional telecasts of this nature would have more than likely happened had the crew had the time to conduct further television operations.

The popularity of an inside peek into the astronauts' activities would have certainly piqued the general public's interest, as was clearly the case on later Apollo flights. Perhaps the crew would have conducted a real-time guided tour of the spacecraft for the benefit of viewers watching from the confines of their own homes. However, other than the above listed activities, no other concrete plans for television transmissions were listed on either the Preliminary or Final Flight Plan.

The Apollo 1 crew accessing the TV camera from its mounted position facing the crew in their seats.

Sadly, all these detailed procedures would remain speculation. On 27 January, 1967 the crew of Grissom, White and Chaffee perished when an outbreak of fire engulfed the entire spacecraft during a routine launch test. The flight of Apollo 1 never actually flew, and the ensuing investigation threatened to derail the United States' efforts to land on the moon within the time challenged by Kennedy. It would not be until the end of 1968 when earth-bound audiences would catch a glimpse of astronauts at work during a space mission. Ironically, the tragic loss of life may have well secured a guaranteed place for the TV camera on all the following flights as previous deficiencies in crew safety fell under the spotlight during the accident investigation.

Following the Apollo 1 fire tragedy, safety factors governing mission redesign saw the use of television within the CSM reinstated in May, 1967 stipulating the monitoring of all three crew members during ground testing and eventual spacecraft launch. Research into the use of the Westinghouse lunar camera for this purpose revealed that heavy bracketing was required to secure the camera in the mandatory position while also shielding it from the shock and vibration of launch. Furthermore, the available lenses for the lunar surface camera could not accommodate the field of view required to monitor all the astronauts together in their seated positions and so it was deemed impractical to use the Westinghouse camera for this function. However, as the already built RCA cameras were smaller and compatible with the Block II electrical power system, NASA opted to refurbish these cameras with the necessary wide-angle lenses and implement them into the CSM cabin, again awarding RCA the contract to carry out the crucial camera upgrades.

The Block II spacecraft program brought about significant changes to the TV requirements and because of the delays brought about by the investigations into the Apollo 1 fire, there was unfortunately no immediate need for the RCA-built cameras. However, they were used in benchmark testing and aircraft missions that flew over ground tracking stations to appraise the air-to-ground radio frequency signal configuration. After a short-lived testing phase the RCA cameras were placed in bonded storage while efforts were concentrated on the lunar surface cameras. It would not be until 1967 that use of TV cameras inside the CSM would be reinstated!

As the Block II CSM was to journey to the moon with the LM attached (the spacecraft now had the all-important docking adapter), NASA imposed a requirement for a single camera which would operate inside both spacecraft and also on the lunar surface. Harking back to the earlier STG suggestion of being able to observe the astronauts live on the moon conducting scientific duties, the cameras had to provide good picture resolution and of course be able to withstand the harsh lunar environment. While the RCA camera was more robust for ordinary lighting conditions, the classified Westinghouse SEC (Secondary Electron Conduction) tube was better suited for operation in extreme low light. This included operation during the extremes of both lunar day and lunar night—in particular low-light scenes occurring in the shaded areas near the LM or those lit solely by earthshine. As a result NASA dealt directly with Westinghouse for a black-and-white camera for use on the lunar surface and in October 1964 officially awarded them a contract for development of this type of camera.

Extreme weight reduction agendas during progression of the Block II program also brought about a significant change in the TV camera placement during the projected lunar missions. The camera was removed from the CSM altogether and placed inside the LM. However, additional weight revisions of the LM ascent stage saw a further repositioning of the camera to the descent stage location of the Modular Equipment Storage Assembly (MESA) which was completely outside of the spacecraft. The camera was to be positioned so that it pointed directly at the LM ladder when the MESA was opened via a lanyard system. The pre-planned position resulted in a clear view of the astronaut descending the ladder and onto the surface of the moon. This setup was ultimately the one used to capture Neil Armstrong's historic step onto the lunar surface on Apollo 11 on 20 July, 1969.

CHAPTER 5. APOLLO 7

"It [the TV camera] was an electrical circuit, and I had not forgotten that an electrical short had resulted in the loss of the Apollo 1 crew."

Wally Schirra.

In the wake of the investigations and subsequent program delays after the tragedy of the Apollo 1 fire in January, 1967, Apollo 7 became the make-or-break flight upon which the future of the planned lunar landings depended. It also, by way of the fire, became the first manned spaceflight of the Apollo program. Watchful eyes scrutinized anything that was electrically powered, and that included the TV camera planned to be used on the flight. Along with an all-out testing of the Command Service Module, real-time television from Low Earth Orbit was to be transmitted from the spacecraft, similar to the intended downlinks which had never eventuated for Apollo 1. Rather than involve three ground stations for reception of the TV signal, as was hoped for on Apollo 1, television pictures on this mission would be transmitted to only two U.S. based ground stations. Commander Wally Schirra noted, "For the first time in the space program we were equipped with an onboard TV camera capable of transmitting to Houston coverage of our mission. With the cooperation of the networks, the first live-from-space television show would be beamed to America and the world."

Design changes in the Block II spacecraft had incorporated an extra S-band transmitter which was fully capable of processing the entire bandwidth of standard broadcast television signals. However, the break-neck pace to launch the flight on time, in conjunction with the lack of unanimous support for the actual TV camera in the first place, resulted in the full resolution TV option being discarded on the upcoming flight. As the RCA slow scan camera had been fully flight tested, and the two ground stations had the appropriate television scan converters up and running, it was decided to use it despite the controversy surrounding its non mission-critical status. Like its forbearer planned for Apollo 1, the compact black-and-white camera operated at 10 frames per second, with 320 lines of resolution, which fell within the originally planned bandwidth limitations imposed on the television system.

The ground stations would then convert the slow scan signal to the broadcast standard of 30 frame-per-seconds, black-and-white NTSC, suitable for relay to Houston, and which would be subsequently sent on for re-broadcast to commercial television stations. Two lenses were included: one wide angle 160 degree lens, and one 9 degree angle lens. The main location for the camera would be locked down in front of the crew while seated in their couches; although the camera could be hand-held, should the astronauts desire that option. This allowed the camera to be taken right up to the command module rendezvous window, or brought to other locations of the spacecraft depending on what the crew decided was worth showing to the TV audience back on earth.

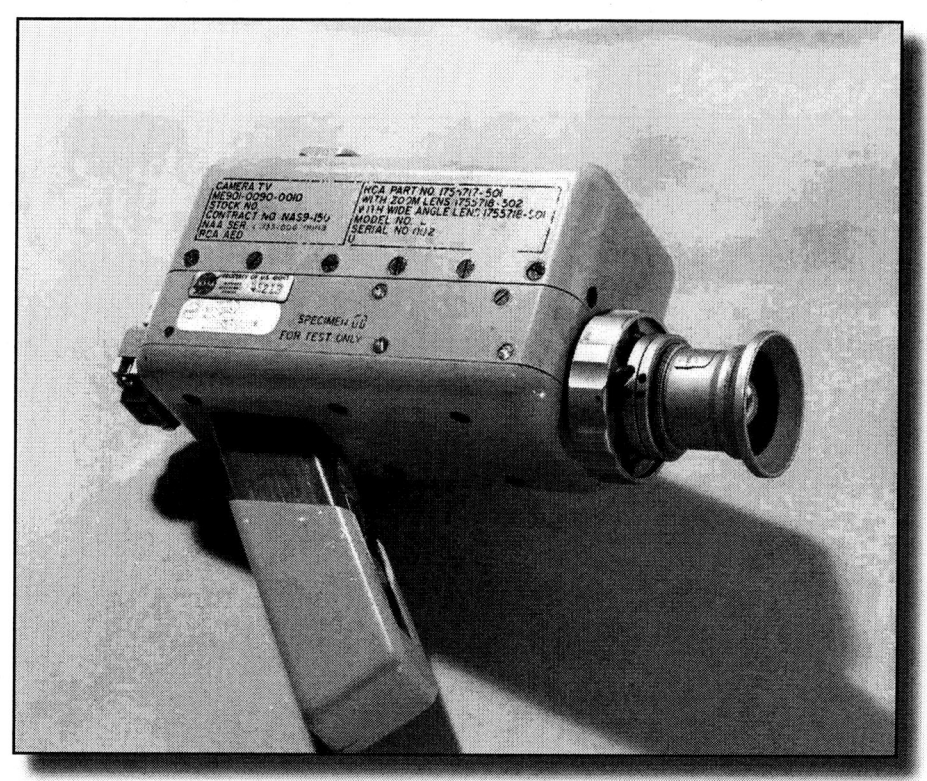

The modified RCA camera now flight qualified for the Apollo 7 mission.
(NASA Photo S68-49009)

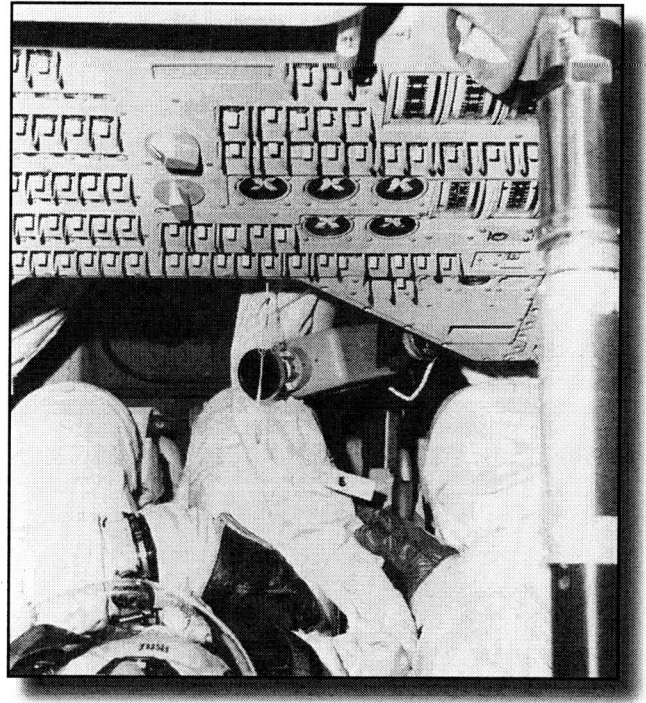

The launch stowage position of the TV camera on Apollo 7.
(NASA Photo 68-H-680)

A NASA press release for the spaceflight advised network TV stations interested in broadcasting the video feed from Apollo 7, that due to the very weak signal at the receiving stations, in conjunction with the low resolution and slow frame rate of the camera, the TV images were expected to be fuzzy and low in contrast. However, although the live nature of the incredible images would set the stage for a very interesting set of transmissions from an orbiting manned spacecraft, the very future of the camera on the flight was in jeopardy.

Schirra explained, "Then there was the weight reduction program. At one point, even our survival kit was affected. Our fresh-from-salt-water pills were cut in half to save eight or ten ounces. We were betting our lives. If we got to the point where it was a matter of survival, what good is a 5-pound TV camera?" It is certainly clear from numerous similar arguments which were made by respective departments at NASA that once again, the engineering and human survival requirement of the missions had placed the use of TV on manned spaceflight at an undue disadvantage. There was no way it could be justified from that perspective. To be fair, Schirra surely felt the pressure of conducting a successful mission and his reluctance for operating yet another electrical device is quite understandable. The shadow of Apollo 1, which took the lives of his comrades and friends, bore heavily down upon him. In today's world of being able to watch anything newsworthy live on television, it may seem a ludicrous concept, but serious debate ensued in NASA review meetings for the mission on whether television should be broadcast at all, and the reluctant decision to carry the 4 1/2-pound camera was not made until shortly prior to the October, 1968 flight.

In a surprise move from General Sam C. Phillips in Washington, the use of television was salvaged via his directive, which effectively over-ruled the earlier decision by NASA to not include a TV camera. Sam Russell, who later worked on the remote controlled lunar television system, recalls that prior to the flight the TV industry placed a lot of subtle pressure upon Sam Phillips. In a May 16, 1969 TV Guide article, "The Raging Space-Shoot Controversy" magazine staff writer David Lachenbruch asked, "Who got to Phillips? "There was no direct communication," said one network producer, "but little items started to appear on the news wires and in the papers referring to the elimination of television. Some pressure started to build up.""

An April, 10, 1968 memorandum from Sam Phillips revealed to all concerned parties that a decision had been reached concerning the television system on Apollo 7. He also outlined several ground rules related to the incorporation and use of the TV camera which would be adhered to throughout the flight. The camera would be installed at the Kennedy Space Center with no impact on the launch schedule. Despite numerous ground tests which had subjected the TV camera to rigorous vibrations, the camera would not be used during the launch phase of the mission. Because of increased safety concerns brought about by the Apollo 1 fire, the camera would have a mounting bracket which would permit observation of the crew during pre-launch hazard tests, in addition to a convenient location for watching the crew in-flight. The camera would also be capable of being hand-held for observations out of the spacecraft window during the mission. Finally, and most significantly for the astronauts, the camera would not be specified on the flight planning timeline, but would be integrated into the mission when the time and opportunity would permit its use.

Lachenbruch further relates the manner in which the TV camera was not overtly introduced into the mission flight plan by stating, "…there was just a list of pencilled instructions inside the spacecraft. Apollo 7's commander, Navy Captain Wally M. Schirra, who ironically became live TV's first astronaut star, had never concealed his opposition to television."

Wally Schirra explained his reasons for not supporting the decision to carry the TV camera on his flight, "My first argument was that we're going to be too busy to provide the world with a vicarious thrill…Whatever goes into the craft I look at down to the last nut and bolt. A television camera is a gee-whiz thing… with no real technical function. On top of that, this one appeared to be inferior Television, like anything else in flight, [it] must pass scrutiny, and this camera wasn't demonstrated as good as it turned out to be."

The haste to secure the TV camera on the flight and subsequently have the crew make a series of test television transmissions had taken them by surprise. Schirra recounts, "The crew wasn't informed prior to launch that a TV was scheduled." He was uncompromising in the fact that if any aspect of the mission had required their full concentration, there was absolutely no way any television broadcast would have occurred. For Schirra, his paramount concern was not how pretty to make the mission to the general public, but how triumphantly he could conduct the Apollo mission to a wary Congress which, following the catastrophe of Apollo 1 had started to doubt the necessity of funding lunar missions.

Apollo 7 launched on October 11, 1968, as scheduled, and carried with it the refurbished RCA TV camera with Wally Schirra as Commander, Donn Eisele as the Command Module Pilot, and Walter Cunningham as the Lunar Module Pilot, even though no actual LM flew on the mission. Television history was about to be made on the first manned mission of Apollo, but not without a certain amount of resistance from the crew. To further compound the defiance to operate what was seen as an unnecessary piece of equipment on the flight, each of the crew members caught a head cold which, in the zero gravity environment caused them all a great deal of discomfort and adversely affected their moods.

The strain to succeed and thus revitalize the goal of a lunar landing was amplified by the head-colds which severely shortened the patience of all crew members when asked to add more items to their already tightly-packed flight plan. In an apparent mutiny against mission control, Wally Schirra made the steadfast decision to adhere to the original plans and not let what he perceived as trivial additions jeopardise the mission. Shortly after insertion into orbit, the S-IVB stage separated from the CSM, the Apollo 7 crew performed a simulated docking with the S-IVB stage, coming to within 1.2 meters of the spent stage of the rocket. Prior to the rendezvous, Mission Control attempted to relay a new set of procedures which involved powering up the television camera. The crew openly objected to this and a few terse words were exchanged over the communications channel:

SCHIRRA: "Roger. You've added two burns to this flight schedule and you've added a urine water dump; and we have a new vehicle up here, and I can tell you at this point TV will be delayed without any further discussion until after the rendezvous."

CAPCOM: "Roger. Copy."

SCHIRRA: "Roger."

CAPCOM: "Apollo 7 - This is CAPCOM number 1.[SLAYTON]"

SCHIRRA: "Roger."

CAPCOM: "All we've agreed to do on this is flip it on."

SCHIRRA: "... with two commanders, Apollo 7"

CAPCOM: "All we have agreed to on this particular pass is to flip the switch on. No other activity associated with TV; I think we are still obligated to do that."

SCHIRRA: "We do not have the equipment out; we have not had an opportunity to follow setting; we have not eaten at this point. At this point, I have a cold. I refuse to foul up our time lines in this way."

Following the uncompromising stance of the Apollo 7 crew, no television was transmitted from the spacecraft until two days later. Schirra noted, "We're still worried about whether this is a safe spacecraft or not." And we had even gotten to the point where they were going to shave all our hair off in case there was a fire. And why am I going to start running a TV show for somebody if I haven't checked the camera out, all the electrical circuits, piece by piece? Ah-hah, it works. Now we'll show you TV."

With a significant time delay which had ruffled the feathers of TV networks in the United States, the first live television from outer space occurred at 71:43 hours GET. The broadcast was a tour of the Command Module, with the crew giving detailed descriptions of experiments and procedures carried on the mission up to that point. Curiously, once the pressure was off them following the successful rendezvous procedure, the crew evidently appeared much more relaxed. Schirra explains his slow change in attitude towards TV, "…we enjoyed television. There was nothing else to do." The main source of fun on the mission for the astronauts was what became affectionately known as the "Wally, Walt and Donn Show" and the antics which were contained therein proved a huge hit with TV networks and their associated earth-bound audiences. Unwittingly the crew of Apollo 7 had accessed the major reason of in-flight TV's mass appeal. It wasn't technical, engineering, safety, nor monitoring of control panels. It was simple entertainment.

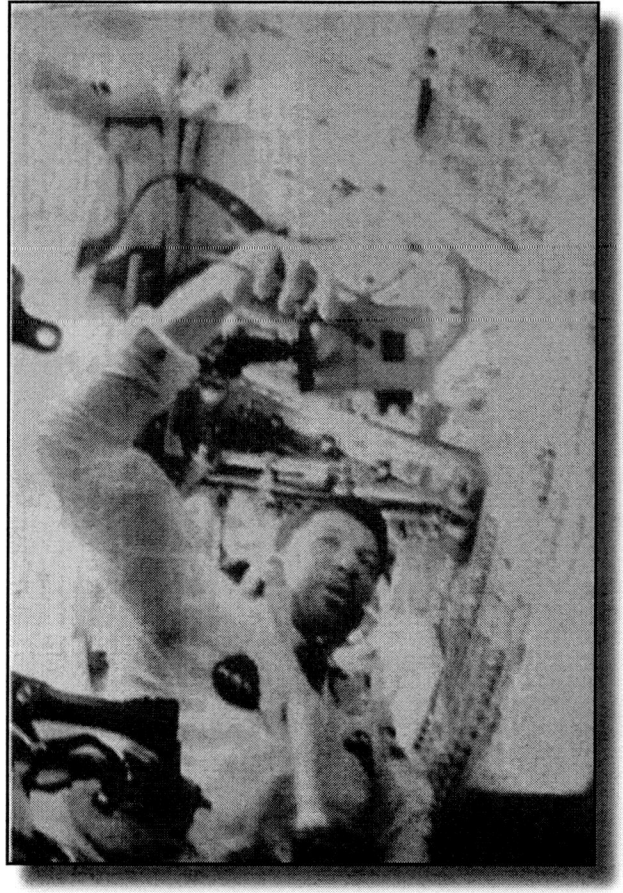

Slow scan monitor scenes from the Apollo 7 transmissions, featuring the famous cue cards and the interior of the Command Module.

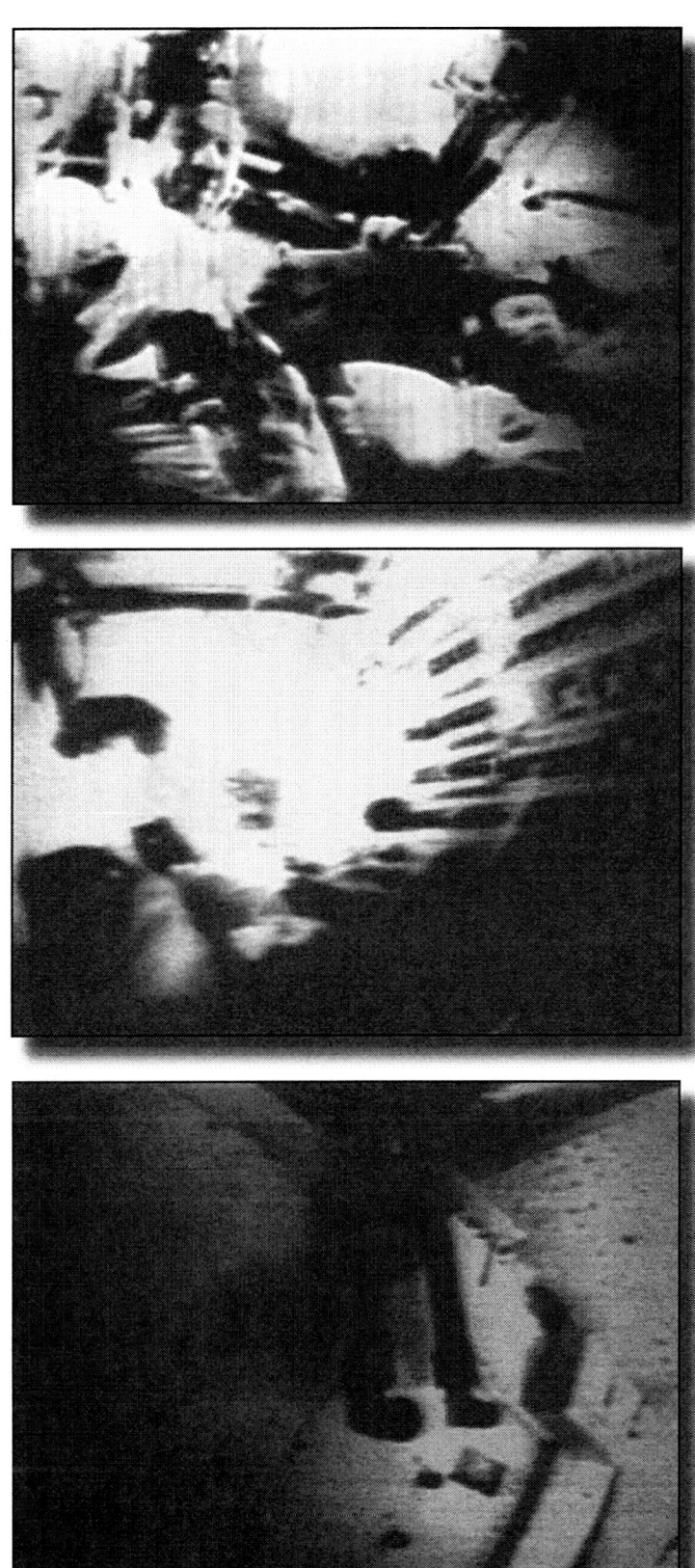

The first transmission started with Wally Schirra and Donn Eisele sitting in the crew chairs holding a now famous hand written cue card with the words, "From the Lovely Apollo Room high atop everything". Capcom "Cecil B de Stafford," as he jokingly referred to himself, "directed" the crew from the ground in the initial phases of the telecast. These directions were solely to get the astronauts in the field of view of the camera, as there was no way for them to monitor what they were shooting. A second cue-card with "Keep those cards and letters coming folks!" was held in front of the camera to the delight of millions of viewers watching in their living rooms. They then provided a view which had never before been seen on the ground, live video of the Florida Peninsula as it passed beneath the spacecraft. Even Tom Stafford commented how obvious orbital motion was on the television feed. Sadly, much too soon the transmission was over. It was possibly the most exciting and riveting footage anyone had seen up to that point. But the visual feast wasn't over, just yet.

The second television transmission occurred nearly one day later, and this time the mood of the crew was even more relaxed, and rather jovial. Highlighting just how temporary the crew's apparent irritation with ground controllers was, Wally Schirra wasted no time in instigating one of his most famous practical jokes. On the air, in full view of the world he used a new set of "cue cards" which were held up to the camera. Several of these cards asked certain participants on the ground if they were a turtle! This was a long running gag which had its origins as far back as World War II amongst fighter pilots. The question, "Are you a turtle?" was always supposed to be answered with "You bet your sweet ass I am!" If for whatever reason the person questioned was unable to respond, they were obliged to buy the next round of drinks. Deke Slayton and Paul Haney were the victims of Schirra's mischievous teasing evident in the conversation between the spacecraft and Mission Control.

CAPCOM: "Just a minute, Wally. Let's see. Oh, it's a little message to Deke Slayton. A little bit closer, Wally. Kind of looks like something about "Are you a, are you a –""

SCHIRRA: "That's right."

CAPCOM: "Looks like it says "Are you a turtle Deke Slayton?""

SCHIRRA: "That's right."

EISELE: "You get A for reading today, Jack"

CAPCOM: "Here comes another one. Wait, oh, that-a-way, that's the way to turn it. It says, "Paul Haney, are you a turtle?""

CUNNINGHAM: "You'll get a gold star; perfect score!"

CAPCOM: "And there is no reply from Paul Haney there."

EISELE: "You mean he's speechless?"

CAPCOM: "Wally, this is Gene. Deke just called in, and we've got your answer, and we've got it recorded for your return."

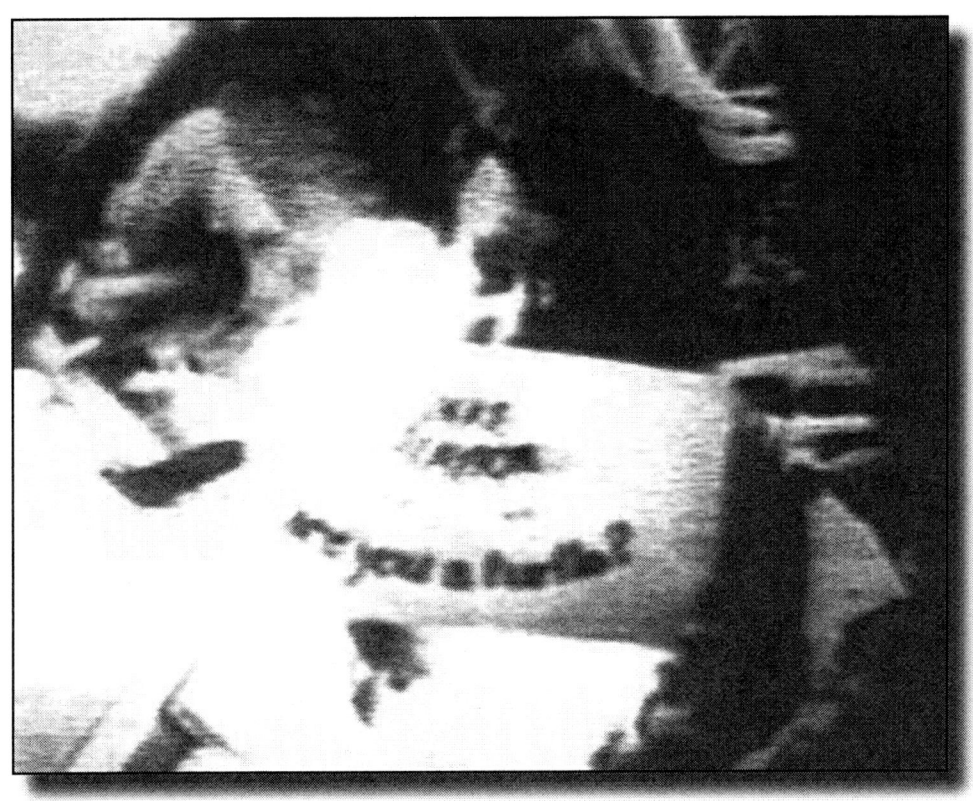

Slow scan monitor picture of Wally Schirra holding the cue card for Deke Slayton asking "are You A Turtle?"

SCHIRRA: "Roger. Real fine. Have you got Haney's answer yet?"

CAPCOM: "No, Haney isn't talking, Wally. Someone tells me he isn't talking, but just buying."

SCHIRRA: "He is buying. Thank you very much. Very good."

The crew then proceeded to show the navigation system and the lower equipment bay section of the Command Module. While searching for the correct aperture setting, an attempt to use the telephoto lens to shoot out of the spacecraft window was aborted as it was felt the camera was unable to accommodate the brightness of the scene. A water and oxygen experiment were shown to earth-bound viewers amidst the "spaghetti" of television cabling floating in front of the camera before the signal was lost once again. The problem being that as only the two stations, Corpus Christi and Merrit Island were able to receive the TV signals from Apollo 7, the TV downlinks were relatively short in duration. When transmissions started again, yet another day later, more water-in-zero-g experiments were conducted for the audiences around the world. They also were treated to a brief explanation of the spacecraft control panel when the telecast came once again to an all-to-soon conclusion. Although the pictures were sub-standard in relation to commercial television requirements, the ability of the viewers to watch astronauts carry out the designated experiments during the flight contributed greatly to public enthusiasm towards the goal of landing on the moon.

The fourth telecast began by showing a view of the hatch window. Unfortunately on this pass, the image was notably poor. Houston requested a switching of antennas to improve the picture which appears to have boosted performance of the TV signal. During the TV show the camera was held on a bright image for a touch too long and resulted in evident image sensor burn in the TV picture. A more detailed tour of the Command Module highlighted where the suit helmets and umbilical hoses were located when the astronauts were connected to the spacecraft's life support system. A momentary loss of stable picture occurred during the cross from Corpus Christi to Merritt Island ground stations after which Schirra demonstrated the unique zero-g situation of floating objects, in this case a Hassleblad camera followed by a pencil and the 16mm Maurer camera with a wide angle lens before signal was lost once again.

The last two TV shows were very relaxed. The course of the mission had seen notable beards appearing on the crew which was referred to in both broadcasts. Further follow-up tours of the spacecraft wrapped up a very successful mission. With the camera discreetly featuring photographs of the astronauts' wives above the Command Module couches, the final scene from Apollo 7 featured the final cue-card with "As the sun sinks slowly in the west". The highly popular "Wally, Walt and Donn" show was coming to an end. There had never before been anything like it.

Back on the ground the TV shows received from space were the buzz in homes around the world. The zero-g demonstrations especially excited children because nothing like this had ever been seen on television before, let alone as it was happening! If that wasn't enough, the fifth transmission would send them into rapture when the crew hilariously performed military marches while floating in the cabin!

Post-flight analysis on the television transmissions for both the first and second tests revealed that due to weak signal strength the image quality would decay rapidly. As evidenced on the second transmission, when the signal markedly improved so too did the resultant video and the increase in clarity proved it. RCA was certainly proud of the performance of its now highly praised camera which had ironically not impressed the astronauts prior to launch. Slowly but surely the perception of television began to change within the space agency. An advertisement for RCA equipment featured the telecasts made by Apollo 7 and detailed how the company had contributed to, not only manned missions, but spaceflight in general.

Following their mission and the hugely popular transmissions from space seen live around the world, the Apollo 7 crew were presented with an Academy of Television Arts and Sciences Special Trustee Award (Emmy) in 1969. If the world was excited by the television they were privy to watch as it happened, the next mission would take that euphoria to a completely new level. Apollo 8 would not only present live views of the world in perspectives never seen before, it would completely change the way humanity viewed itself.

An RCA Advertisment 1968.
(Courtesy of Lockheed Martin. Used with permission)

The cue card featuring "As the Sun Sinks Slowly in the West…"

TV Transmission	GET hh:mm	GMT hh:mm	GMT Date
1st TV Transmission START	71:43:00	14:45	10/14/1968
1st TV Transmission END	71:50:00	14:52	
2nd TV Transmission START	95:25:00	14:27	15.10:1968
2nd TV Transmission END	95:36:00	14:38	
3rd TV Transmission START	119:08:00	14:10	10/16/1968
3rd TV Transmission END	119:18:00	14:20	
4th TV Transmission START	141:11:00	12:13	10/17/1968
4th TV Transmission END	141:27:00	12:29	
5th TV Transmission START	189:04:00	12:06	10/19/1968
5th TV Transmission END*	~189:12	~12:14	
6th TV Transmission START	213:10:00	12:12	10/20/1968
6th TV Transmission END*	~213:20	~12:22	
7th TV Transmission START	236:18:00	11:20	10/21/1968
7th TV Transmission END	236:29:00	11:31	

TV timeline from the Apollo 7 mission. *End times of the 5th and 6th Transmission times are approximate as the NASA timelines are incomplete. Given end times are made from observing the actual downlink videotape.

CHAPTER 6. APOLLO 8

"I didn't want to take the damn television camera with me."

Frank Borman.

1968 is a year that often is mentioned as one that people would rather forget. Turmoil the world over seemed to dominate the news headlines, and as the year came to a close there was very little to feel excited about. Coming at the end of this particularly unpleasant year which predominantly seemed to be filled with bloodshed, assassinations and human misery, the flight of Apollo 8 would provide people around the world with something sorely lacking in 1968. Hope and wonder - which came directly to them via television. The flight would also catapult NASA's readiness for a lunar landing, bringing the goal of Kennedy well within reach by the end of the decade.

Apollo 8 was arguably the biggest milestone of the entire Apollo program, for without its success, the landing of Apollo 11 may never have happened, nor any other manned lunar mission for that matter. The hugely successful flight was born out of a bold decision, made in the face of significant delays which threatened to push a moon landing out of the 1960s, thereby missing President Kennedy's deadline.

Grumman's delivery of the Lunar Module was pushed into the first quarter of 1969, which would miss the flight of Apollo 8 that had planned to test it in Earth Orbit. George Low, manager of the Apollo Spacecraft Program Office was forced to confront a hefty dilemma. Without a Lunar Module, Apollo 8 would simply be a repeat of the previous flight of Apollo 7, which had performed to "110%" as the post-flight debriefings had confirmed. At the same time the CIA had relayed information which reported that the Soviet Union would attempt a manned spaceflight to the moon using its Zond spacecraft before the end of 1968. Low's solution was both simple and brilliant. Send Apollo 8 on a circumlunar mission without the Lunar Module. If successful, the Americans would be the first to reach lunar orbit, and would also clear the way to beat the Russians to an actual landing.

After some decisive persuasion from his colleagues, James Webb, NASA Administrator, finally also saw the benefit of changing the mission and upgraded it to a C-Prime Lunar Orbit flight. The decision was withheld from public knowledge until after the Apollo 7 flight; being made public a mere 40 days prior to the flight. Apollo 8 was going to fly to the moon.

The Final Photographic and TV Operations Plan Apollo 8 from November, 1968 briefly outlined the purpose of in-flight television transmissions stating they would, "provide information which will assist in evaluation of the potential usefulness of TV from lunar distance." The report also placed high importance on the crew obtaining views of the moon's surface and the earth as seen from lunar distances, as well as interior views of the crew during the flight. Only two tracking stations, Madrid and Goldstone would be capable of receiving the TV signal through the whole mission. However, unlike the previous Low Earth Orbit transmissions, Apollo 8 would maintain a near-fixed line of sight with the two stations which would allow lengthy use of the TV broadcasts using the steerable S-band antenna on the Command Module. The mission of Apollo 8 was to use television equipment which was identical to that flown on Apollo 7. Like its predecessor, the Command Module had the equipment necessary to allow a full bandwidth TV system to be used, however time restraints prevented it from being upgraded with the necessary cameras. The flight tested RCA black-and-white camera was to be used with both a supplied wide angle and telephoto lens, and a list of planned transmission times, along with the available tracking station, was supplied (which interestingly was adhered to rather closely during the actual flight).

There was, however one problem concerning the proposal of television on the flight. Frank Borman, mission commander was quite vocal about his desire not to take the TV camera because, like many astronauts, he felt the use of television would serve absolutely no purpose towards the fulfilment of mission objectives. "I said 'no' a lot, and the nice thing about it was that NASA gave the commander enough prerogative that they backed him up. I was overruled on one thing and that was because management was a lot smarter than I was. I didn't want to take the damn television camera with me. And they said, 'Let's take it,' and they were right. ... It turned out to be so important because we could share what we saw with the world. It weighed 12 pounds. We were cutting out everything, even down to the extra meals, which weighed 16 ounces or something like that. But I was very short sighted there, and NASA was right."

Tracking Station	Planned Signal Acquisition GET hh:mm	Planned Transmission Duration
Goldstone	31:00:00	15
Goldstone	55:00:00	15
Madrid	72:10:00	15
Goldstone	85:37:00	15
Madrid	98:00:00	15
Goldstone	128:00:00	15

Tracking station plans for Apollo 8.

Apollo 8 launched on December 21, 1968 with Commander Frank Borman, Command Module Pilot Jim Lovell, and Lunar Module Pilot Bill Anders, although, like Apollo 7, there was no actual LM on this flight either. For two hours and thirty-eight minutes after launch the crew worked feverishly to ensure the rocket was ready for Trans Lunar Injection. They took photographs of the spent S-IVB stage before being given the command to officially begin their journey to the moon. Given the nature of the flight (i.e. this was the first attempt to orbit the moon) there was no scheduled TV at any time prior to TLI, the point of the mission where the spacecraft would be officially sent on its way to the moon.

Approximately 25 minutes prior to the officially scheduled broadcast Borman had requested a brief test of the television system. Unfortunately as the high gain antenna was about to lose its favourable position shortly after the request was made, the test was not conducted. Capcom Ken Mattingly also explained that the signal could not have been fed through to Houston live, so the idea was scrubbed. However, as planned, at 31:10 GET the crew of Apollo 8 turned on the circuits to begin transmitting television from the spacecraft while it headed toward the moon. The first view audiences saw was that of Frank Borman sitting in his couch with his hand on the Rotational Hand Controller guiding the spacecraft to allow the camera a view of the earth. The feet of Jim Lovell who was working in the Lower Equipment Bay could be seen jutting out, and the dextrous camera work by Bill Anders provided surprisingly clear views of the Command Module. The unique transmitting location was made apparent by a floating plastic bag which happily moved around the camera's field of view while the astronauts described certain aspects of spaceflight. Whenever anything bright was held in shot for too long, however "burning" of the image on the camera tube occurred - a problem that if not paid attention to could potentially ruin the camera.

Frank Borman on the first transmission from Apollo 8.

An attempt was made to attach the telephoto lens to the TV camera and shoot the earth through the spacecraft windows. The moment Anders attempted to put the earth in shot, Houston saw a set of grey luminance bars, much like the standard color bars seen in studios around the world, except of course they were without any color. Something was definitely wrong! In a frantic attempt to pinpoint the cause, Flight Control quizzed all relevant stations in Mission Control as to their potential understanding of why no image was being seen on the ground. The Electrical, Environmental and Consumables Manager (EECOM) finally reported that Goldstone was receiving an extremely high white level from the spacecraft TV image, meaning that something on the camera was not functioning properly.

Numerous attempts were made to close the iris of the camera down to allow less light onto the picture tube, as well as switching the automatic light control off on the camera, but it made no difference. At this point the crew replaced the telephoto lens with the wide angle lens and produced a TV image of a faint circular object in the window. That object was the earth receding into the background as Apollo 8 neared the moon. Mission Control were concerned that prolonged use of the telephoto lens would indeed burn the pickup tube inside the camera rendering it useless, and so the crew was advised just to continue shooting inside the capsule. In an effort to dispel concerns of the ground staff looking at the problem, Borman announced the earth had a reading of 320 lumens according to his Minolta light meter. Reflecting on the personal aspect reflected in the telecasts, Bill Anders would comment about them stating that, "It's a means to bring people with us on these trips. It's good for them and it's great for us, because it gets the people enthusiastic – it's their program. TV brought it home, made it a little more personal for them."

The greyscale bars seen when the camera failed to return an acceptable picture.

Just before Anders turned off the camera, Lovell cheekily grinned and announced, "Happy birthday, Mother!" Frank Borman later noted the public opinion of the telecast by adding, "We all hammed it up a bit, but after we got back I heard that when CBS interrupted the pro football playoff game between the Vikings and Colts for our brief broadcast Sunday afternoon, the network had been swamped with protesting calls. Maybe we should have thrown a football around."

The next day's transmission started with some concern at Mission Control. The television circuit was operational but all the ground control personnel could see was a black screen, which prompted a fury of diagnostic checks to see what the cause of the problem may have been. However their fears were soon put to rest when a sharp bright image moved into shot filling nearly a third of the screen. For the first time in history the people of the earth were looking back down upon themselves. In the time between the first and second transmissions a number of specialists in Houston had worked to find a resolution to the bright image problem which had kept the telephoto from bringing views of the earth as planned.

The view of the earth as shown to eager TV viewers around the world.

Two hours prior to this scheduled TV broadcast Borman had noticed the spacecraft would offer a superb view of the earth and began discussing the upcoming broadcast with Houston.

COLLINS: "Roger, Frank. I've got a lot of talking to do regarding TV cameras and brackets and what not. I would like to start in on it whenever you are ready to talk about it. Okay. First a question. Are you planning to show us TV pictures of the Earth today?"

BORMAN: "Well, that is what we wanted to do. It seems that would be the most interesting thing we can show you, but we - you know, we had trouble with the lens."

COLLINS: "Well, okay, that's good. All this procedure that I am going to give to you here is relative to what we hope are fixes to the lens and for looking out your rendezvous window at the Earth, and all the gimbal angles and all that good stuff is based toward looking out the window at the Earth rather than at the Moon. Over."

BORMAN: "Roger."

COLLINS: "Okay. First, unstow the red filter, the polarizing filter, the red and blue filter holder, and some tape. Over."

BORMAN: "Okay. Let me write this down."

COLLINS: "Roger. I'd suggest that. I've got a whole page full."

BORMAN: "Okay."

COLLINS: "Alright. Tape the red filter to the telephoto lens. That red filter is the 25A red filter, not the one that is in the red and blue filter slider."

BORMAN: "Roger."

COLLINS: "Attach telephoto lens to the camera."

BORMAN: "Okay. We can figure out how to do that. Roger."

COLLINS: "Ensure that the automatic light control, the ALC switch on the camera, is in the In position. Over."

BORMAN: "ALC In. Roger."

COLLINS: "Roger. Attach camera to the adjustable TV bracket and attach the bracket to the TV mounting point on the Commander's side of the hatch to point out rendezvous window number 2."

BORMAN: "Roger."

By using the red filter it was hoped that not only just the blueness, but the overall brightness of the earth would be diminished thereby producing an image which would not be too intense for the TV camera to process. The entire broadcast had now been predetermined to only show the earth and so all the position plans for the spacecraft in order to present an uninterrupted view of the earth took this into account. As the television images were being sent to earth the astronauts provided descriptive commentary by using binoculars to observe the earth as well. They, of course, had the benefit of color, which greatly assisted in their determination of what was landmass and what was water. While the images in the transmission were remarkably good, there was a great deal of difficulty in aligning the camera so as to have the earth in the center of the frame. Bill Anders would later note, "For a while they kept telling us, 'Move it up' or 'Move it down'. We kept moving it the wrong way. In space, who knows which direction is up?" Capcom guided the astronauts, who in turn moved the camera in its mounting bracket and also physically repositioned the spacecraft to achieve the required framing of the earth. This lengthy procedure further prompted Anders to remark in a crew discussion shortly after the telecast (and captured by the on board recorder), "Also, I'd like to suggest that if they ever fly one of these TV cameras again, they put a - some kind of a sight on it. Sort of ridiculous to have a 9-degree field-of-view lens with no way of aiming it except for looking down the side or putting some chewing gum on the top."

Regardless of the difficult process in finally getting a decent view of the earth the end result was well worth it. The lasting image of this transmission was one which would inspire artists and poets forever more - an isolated earth sitting in a sea of blackness. As one earth bound news correspondent remarked during the telecast, "lest there be any doubts as to whether the earth is round, and there are still those who belong to organisations who don't believe it, there it is – live on television."

The third telecast was a close-up view of the lunar surface while the spacecraft was in orbit. Definition was exceptionally clear with unmistakable vistas of the moon passing underneath the capsule. Craters, rills and lunar mare were all visible as Anders described the names of the surface features when they came into view. The compact RCA camera was providing earth-bound viewers their most detailed look at the moon ever as the spacecraft flew over it.

The simplicity of the camera did not overshadow the impact of the images sent back to earth. As Jim Lovell later recalled, "And at the same time, we had this sort of now rudimentary TV camera, black-and white camera, that was pointing out the window watching the craters go by and slowly slipping into daylight". However, as enduring as these views of the lunar surface from 60 miles up would become in the human psyche, it would be the next transmission scheduled to occur on Christmas Day, 1968 which would become iconic of the nation's space program.

The opening scene of the 4th telecast featured a shot taken out of the spacecraft window and featured a small disc-shaped object which was the earth. Without the telephoto lens the planet seemed tiny and its apparent size emphasized just how far away the crew of Apollo 8 was from it. They then moved the camera to a fixed location which looked down on the passing lunar landscape, similar to the views afforded them the day before from lunar orbit. After discussing many of the features which were clearly visible on the TV image, the crew proceeded to read from the Book of Genesis.

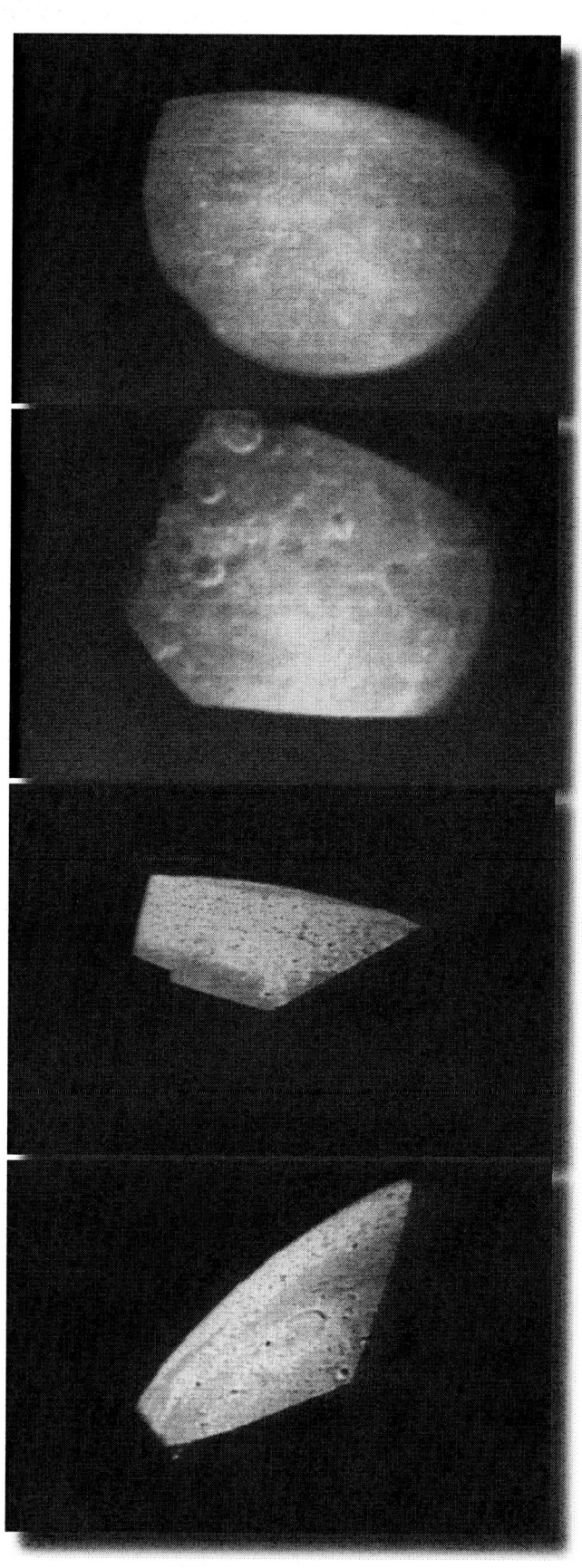

The first live lunar views from Apollo 8.

ANDERS: "We are now approaching lunar sunrise, and for all the people back on Earth, the crew of Apollo 8 has a message that we would like to send to you. In the beginning, God created the Heaven and the Earth. And the Earth was without form and void, and darkness was upon the face of the deep. And the spirit of God moved upon the face of the waters, and God said, "Let there be light." And there was light. And God saw the light, that it was good, and God divided the light from the darkness."

LOVELL: "And God called the light Day, and the darkness he called Night. And the evening and the morning were the first day. And God said, "Let there be a firmament in the midst of the waters. And let it divide the waters from the waters." And God made the firmament and divided the waters which were under the firmament from the waters which were above the firmament. And it was so. And God called the firmament Heaven. And the evening and the morning were the second day."

BORMAN: "And God said, "Let the waters under the Heavens be gathered together into one place. And let the dry land appear." And it was so. And God called the dry land Earth. And the gathering together of the waters called the seas. And God saw that it was good. And from the crew of Apollo 8, we close with good night, good luck, a Merry Christmas and God bless all of you - all of you on the good Earth."

LOVELL (onboard): "That's it."

BORMAN (onboard): "Don't say anymore now."

ANDERS (onboard): "I just turned it Off. You want it On again?"

BORMAN (onboard): "No. leave it Off. Great! Great!"

ANDERS (onboard): "Off?"

BORMAN (onboard): "Yes."

ANDERS (onboard): "Okay."

LOVELL (onboard): "Camera's Off?"

ANDERS (onboard): "Yes."

BORMAN (onboard): "Hey, how can you beat that? Geeze, we just went into the terminator right in time."

It is obvious from the onboard recordings that Frank Borman was well aware of the significance of this TV show from lunar orbit on Christmas Day. What he was not totally conscious of was the impact the reading would have. Certainly the guidance by NASA to "do something appropriate" had been met a dozen times over. Nevertheless the mission was not over yet. Two more telecasts were scheduled during which the crew demonstrated more aspects of life on Apollo 8. Jim Lovell, dubiously dubbed "The Presidential Advisor on Physical Fitness" performed exercises to exhibit the manner in which the astronauts maintained their stamina in the world of zero-g.

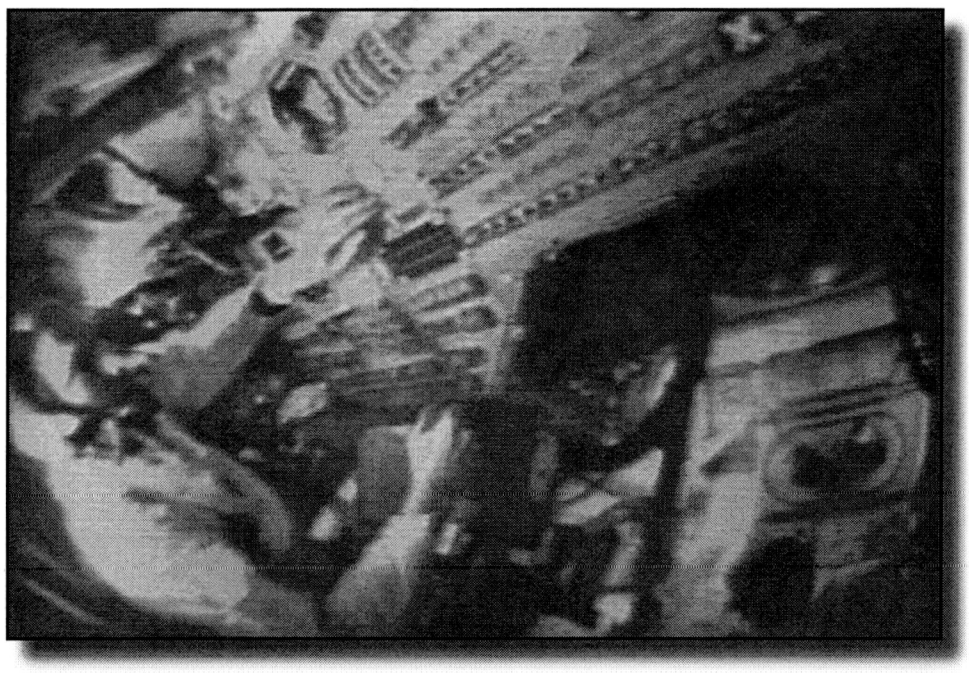

The computer system was briefly featured while Bill Anders was preparing a demonstration of food and orange juice from the food storage area. A rudimentary display of the navigational and optical system was shown before the camera moved in on a very clearly visible mission logo on Lovell's jacket sleeve and ended the TV show. The final transmission featured closer views of the earth as the spacecraft approached for its re-entry into the atmosphere. The crew gave discussions of the visible areas of the earth and mentioned the fact that very shortly TV audiences would be able to watch live television from the surface of the moon. In signing off Borman simply said, "See you on that good earth real soon."

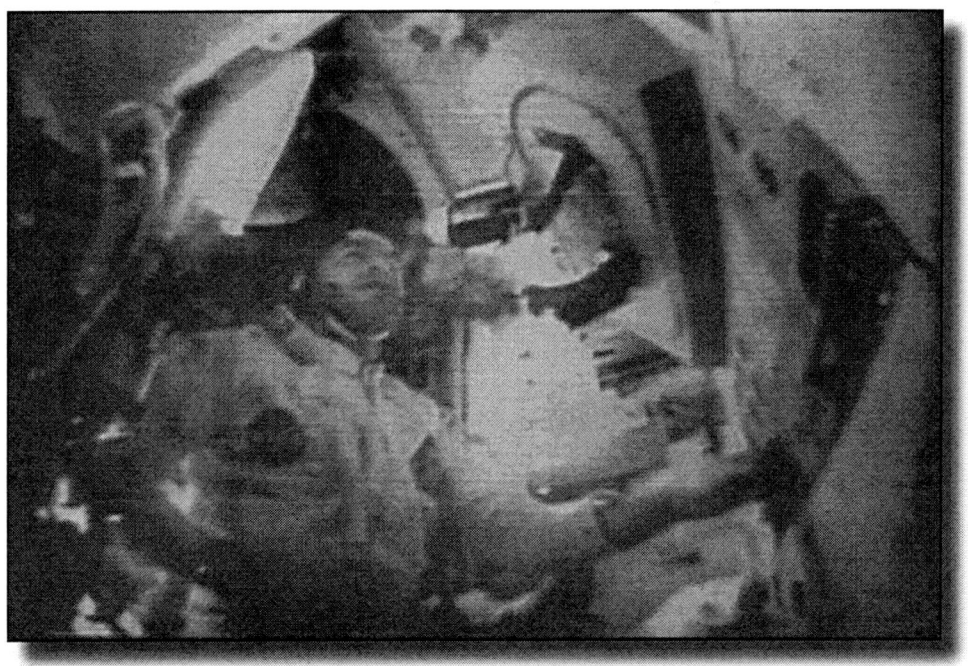

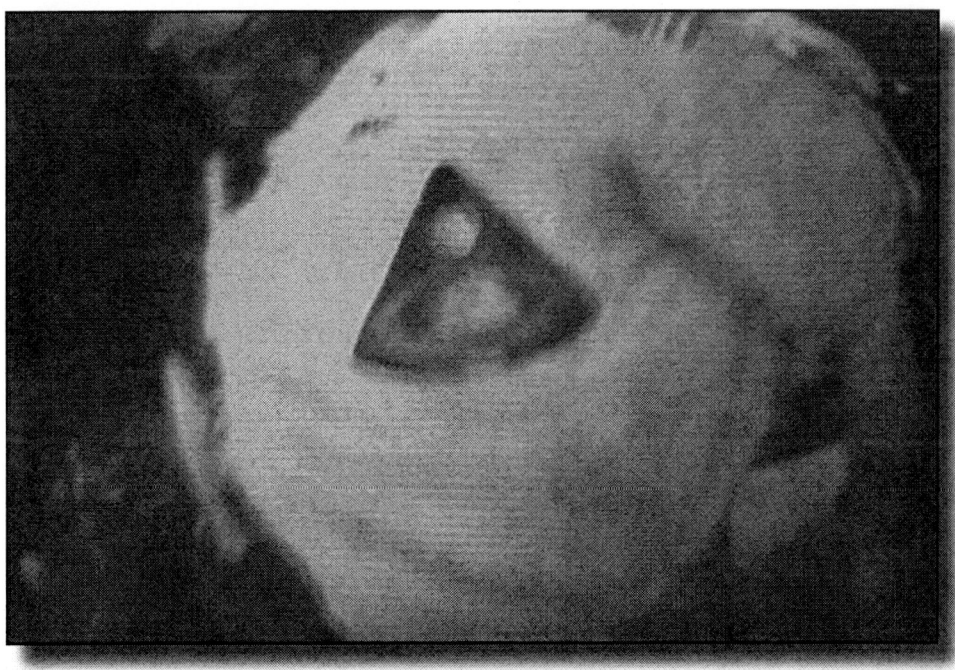

The highly successful mission had set the stage for the first lunar landing. It was a huge gamble by NASA and it had paid off immensely. Public interest and excitement was at a peak. Network ratings for the Apollo 8 telecasts were the largest ever seen for space coverage. Sadly, despite having sent spectacular black-and-white images on two Apollo spaceflights which had helped to change humanity's perception of itself, the small RCA camera would be retired after the flight of Apollo 8. In tribute to the enormous cultural impact the camera had provided to the world via its television pictures, Jim Lovell simply stated, "The only controversy was that we didn't give them enough."

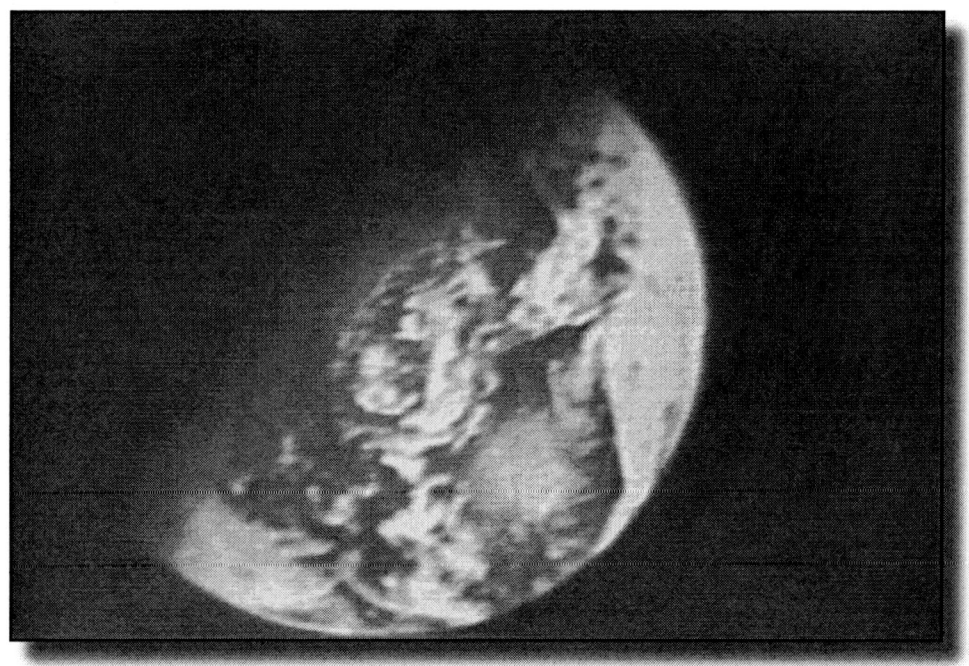

Television Transmission	GET hh:mm	GMT hh:mm	GMT Date
1st TV Transmission START	31:10:00	20:01	12/22/1968
1st TV Transmission END	31:24:00	20:15	
2nd TV Transmission START	55:02:00	19:53	12/23/1968
2nd TV Transmission END	55:28:00	20:19	
3rd TV Transmission START	71:40:00	12:31	12/24/1968
3rd TV Transmission END	71:52:00	12:43	
4th TV Transmission START	85:43:00	2:34	12/25/1968
4th TV Transmission END	86:09:00	3:00	
5th TV Transmission START	104:24:00	21:15	12/25/1968
5th TV Transmission END	105:33:00	21:24	
6th TV Transmission START	127:45:00	20:36	12/26/1968
6th TV Transmission END	128:05:00	20:56	

Apollo 8 Television Transmission Schedule.

CHAPTER 7. APOLLO 9

"We don't have any skits but I hope we don't have any novel or new things!"

Jim McDivitt in response to a question about the planned television events for Apollo 9.

The early months of 1969 reinforced just how little time was remaining in order to fulfil Kennedy's promise of a lunar landing before the decade ended. No longer was the deadline "a few years from now" or "next year". It was a matter of months. Despite the successful flights of Apollo 7 and 8, the pressure was certainly mounting as the end of the year rapidly approached. NASA could not afford any slip-ups which would have potentially set the landing date well into the 1970s. The long anticipated mission of Apollo 9 in March of 1969 was finally ready to test both the Lunar Module and the Command Module in earth orbit. Grumman had finally overcome the numerous delays and setbacks, and had delivered LM-3, the first Lunar Module to be used for manned flight. The primary objectives of the flight were to evaluate manoeuvrability of the lunar module and to establish docked vehicle functions in an earth orbital mission, thereby qualifying the combined spacecraft configuration for lunar flight. Lunar Module operations included a descent engine firing while docked with the Command Module, a complete rendezvous and docking profile, and, with the vehicle unmanned, an ascent engine firing to propellant depletion. Combined spacecraft functions included command module docking with the lunar module, spacecraft separation from the launch vehicle, five service propulsion firings while docked, a docked descent engine firing, and extravehicular crew operations from both the lunar and command modules. Returning to the tradition established during project Mercury, the crew affectionately named their spacecraft, also in part to ease confusion in communications once the two spacecraft had separated. In the case of Apollo 9, the Command Module would be known as "Gumdrop" given its resemblance to one, while the Lunar Module would be named "Spider" for its landing legs and bug-like appearance.

Unlike any other Apollo mission, no camera was to be carried aboard the Command Module. Because Apollo 9 was an earth-orbital mission, the requirement for television from the CM was not included in the flight plan and so no CM TV camera was put on board, although the Westinghouse Lunar Surface Camera was stored inside the LM. It was decided that the camera should be operated from space through the LM communications system before the camera was used on the lunar surface, and so the camera was solely operated from the LM cabin during the two television passes of the mission.

Stan Lebar demonstrates the lunar surface camera which was tested on Apollo 9. (NASA Photo 69-H-152 courtesy of Stan Lebar)

The Westinghouse TV camera was equipped with an 80-degree wide-angle lens and a 35-degree lunar-day lens. Depth of focus for the wide angle lens was from 20 inches to infinity and for the lunar-day lens from 11 feet to infinity. Similar to the RCA black and white camera which had flown on Apollo 7 and Apollo 8, the camera had a 10-frame-per-second scanning mode, with an addition not found on the RCA models: a high resolution mode which would scan at 0.628 frames-per-second. The camera, utilizing the classified SEC tube could televise scenes of low-light intensity making it ideal for the diverse lighting conditions which would be encountered during the flight.

"The TV camera used on Apollo 9 was identical to the TV camera used for the Apollo 11 EVA." noted Stan Lebar. He further explained why the RCA camera was replaced with the Westinghouse lunar surface camera on the mission, "From Apollo 9 on, NASA directed that the WEC Lunar TV camera was to support the TV requirements for the CSM, LM and lunar EVA. With regard to the spacecraft TV, Apollo 9 was used to test the TV transmission from the LM, and the TV converter and hand-off to the broadcast network as would be used during the forthcoming missions, including a lunar EVA. Originally it was planned that Rusty Schweickart would operate the camera from the LM porch in the vacuum environment as would be the case on the lunar surface but…that was scrubbed…"

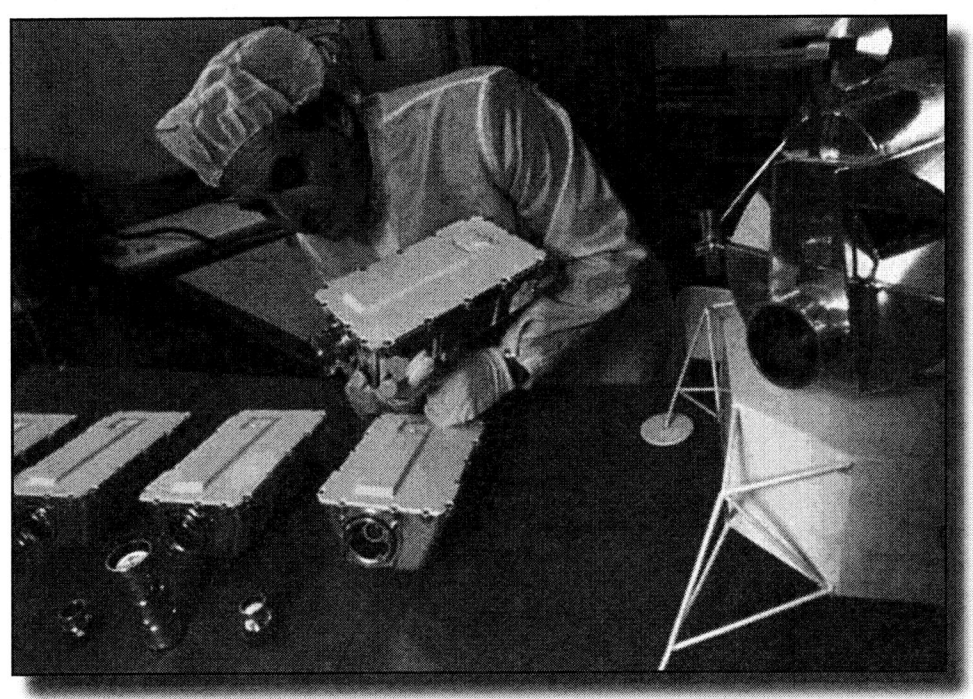

Stan Lebar displaying the several TV cameras. Note the earlier LM design model to the right of the picture.

"Since Apollo 9 was in earth orbit, a slow scan TV converter was located at The Corpus Christie Tracking Station. The slow scan converter used for the Apollo 9 mission (and previously used for Apollo 8) was one that had been designed and built at JSC to support the TV simulations that were conducted at JSC. From Apollo 10 on, only the WEC field sequential color TV was used in the CSM. The Apollo 11 was the last Apollo mission that made use of the slow scan black and white TV on the lunar surface. The TV camera used on Apollo 9 was identical to the TV camera used for the Apollo 11 EVA."

What is astonishing about pre-flight preparation is that even after the hugely successful TV telecasts of the previous manned Apollo flights, the mental impasse of NASA management in seeing television as something more than just an engineering tool had not yet been crossed, although indications had begun to appear which suggest the attitude was indeed slowly changing. Robert Gardiner, Chief of Guidance and Control, issued a memo in which he discussed the scientific benefit of television transmissions during crew activities in the Lunar Module in order to test the camera itself. However, no mention was made at all of the "entertainment" or publicity aspect of utilizing television on the flight. The Gardiner memo listed the following as recommended procedures for the two television transmissions:

"The first TV pass occurs on Rev. 30 around 46:28 GET and lasts approximate 5 minutes. It is desirable that about (3) views be taken: (1) the instrument panels of recognizable areas, (2) the aft stowage area behind the crewmen, and (3) side view of the crewman not holding the camera. During this time the crewman should hold the camera steady and allow about one minute of each view to compensate for motion and light level adjustment."

"The second TV pass occurs at Rev. 48 around 75:00 GET near the end of EVA and lasts approximately 15 minutes (Goldstone to MILA coverage). The CDR connects TV camera and then passes to the EVA crewman. It is desired that the CDR point the camera viewing out the hatch at the EVA crewman for about a minute and then transfer camera to him. The EVA crewman is requested to point the TV camera at general targets indicated by circles on the attached figure (docked CSM and LM). Again it is recommended that the camera be held as steady as possible for approximately one minute to compensate for motion and light level adjustment. If time permits, near termination of transmission, the crewman should attempt a shot of the earth to determine if a meaningful picture can be obtained."

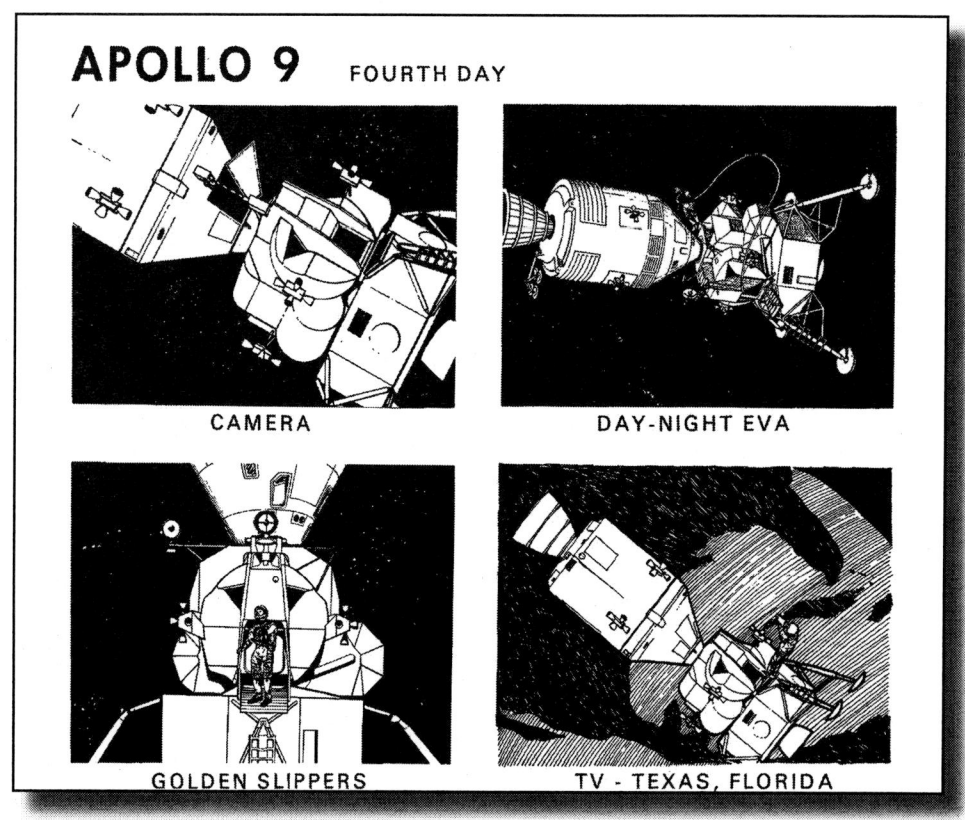

Planned activities on day 4 of the Apollo 9 mission. Note the use of the TV camera for the EVA (From NASA Image S69-19795).

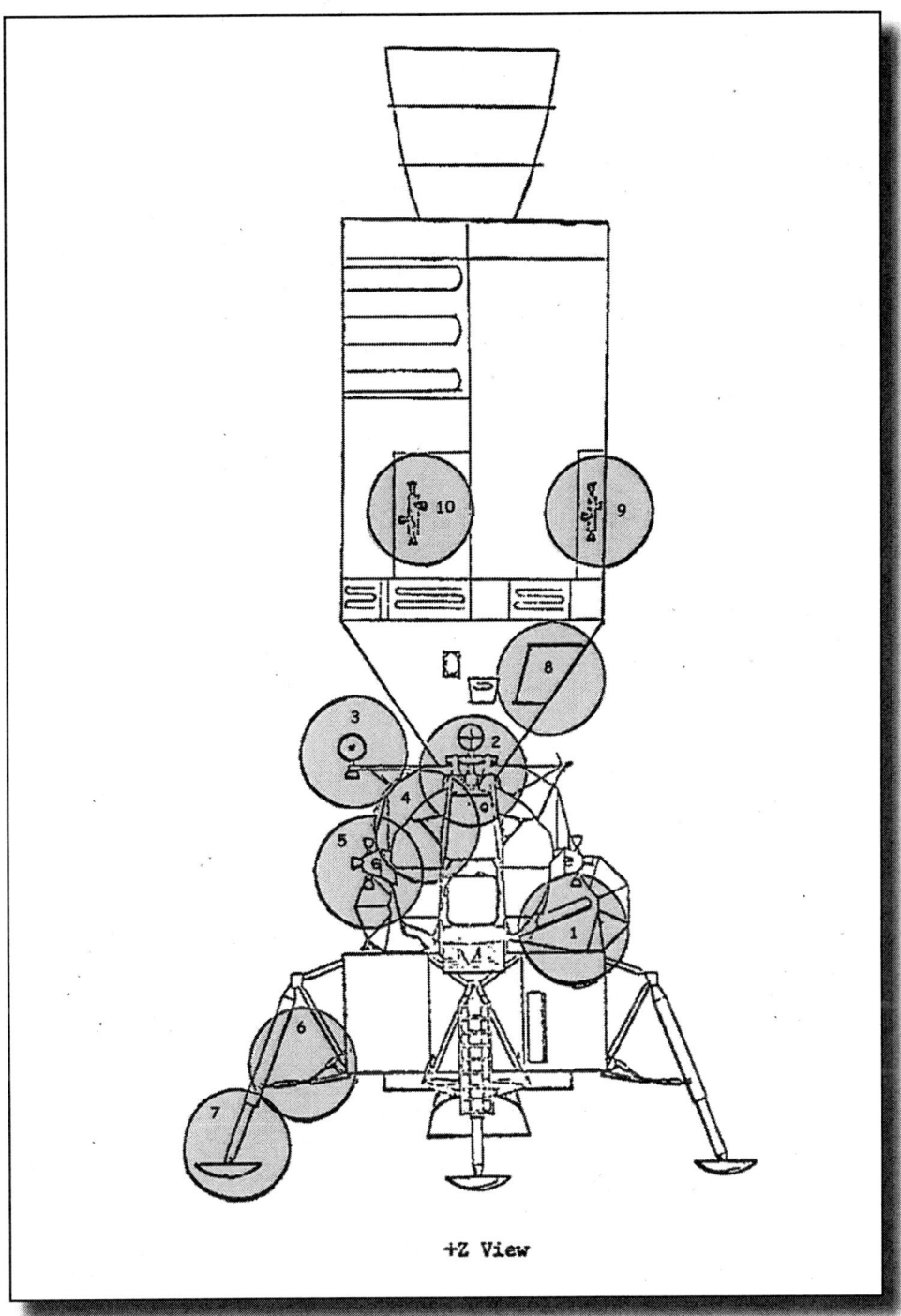

The proposed camera targets (marked in a grey shade) for the aborted EVA TV transmission.

Despite NASA's apparent ambivalence towards the scheduled television shows, TV networks across the United States were excitedly preparing for a live transmission of the planned EVA. Shortly before launch, Schweickart explained to an eager corps of journalists at the Pre-Mission Press Conference, exactly what was planned as far as television coverage was concerned, "We have two TV passes scheduled. The first one will come on the third day – the systems day, and I can almost assure you this is not going to be very glamorous TV. We don't have the capability to mount the camera properly inside the LM; it wasn't designed for that, and we're going to do the best we can. But in essence what we are going to do is put the TV camera in the back of the spacecraft, point it toward the front where we're operating and let it go, and you're not going to hear much on this except what we're doing because as Mr Boone observed we've got a lot to do in there and I would expect you're going to see the normal day in the life of a LM much more than the "Jim, Dave and Rusty Show. The EVA transfer itself will not be [televised], the second time we have a TV pass will occur during the EVA. It's during the second day pass, not the second day, during the EVA. We do have a TV pass at that point and we're still working out the final details but it appears the camera will be mounted on the porch and will be looking up across the docked vehicles, and it will be stationary at that point."

On the day of the launch Walter Cronkite reiterated Schweickart's detailed explanation of planned TV during his CBS coverage of the flight to his television audience, "During this flight there will also be live television from space - the first test of a new Westinghouse television camera; the camera that will be used on the surface of the moon. The previous cameras we have seen, RCA cameras which have worked so well in the past, have been supplanted now by this Westinghouse camera which was the plan all along - the camera which can withstand the low pressurization on the moon. It weighs almost twice what that former camera did."

Apollo 9 launched on the 3rd of March, 1969 with the crew of Commander Jim McDivitt, Command Module Pilot Dave Scott and Lunar Module Pilot Rusty Schweickart. Along with a fully operational Lunar Module, they also carried with them the Westinghouse lunar surface camera stored inside the Lunar Module, as specified in NASA plans prior to the flight. Once the crew had successfully attained orbit, a problem presented itself which could have thrown the entire mission into jeopardy. Rusty Schweickart became violently ill due to space sickness--a condition in which astronauts become sick when suddenly faced with the unsettling environment of zero gravity. The consequences of an astronaut vomiting while wearing his spacesuit in zero-g would without question be fatal. Jim McDivitt, sensing the problem would eventually subside, decided to postpone the EVA while Schweickart acclimatized himself to the new environment in space. Discussions between Houston and the crew prior to the spacewalk reveal that even at that point in the mission, however, television was still intended for at least part of the EVA, albeit with the non-intrusive schedule which would allow the astronauts to achieve the mission requirement of the spacewalk before they presented a TV show for those on the ground, just as Rusty Schweickart had described at the Pre-Mission Press Conference.

SPACECRAFT: "It looks like we're going to have to open the hatch at normal time, leave it open for that daylight pass, close it, configure it for the TV, and when the TV is over then we would leave the LM, come back in the command module, Is that right?"

CAPCOM: "That's right and as a matter of fact, we don't even want the TV to interrupt the transfer. If possible you can, you know, start the transfer early."

SPACECRAFT: "Oh okay I see what you're saying. Your saying we plan to follow normal timeline and when we get to the time to open-the hatches, we do that, leave them open during the first daylight pass, close them up, and then we egress the LM, and tune in the TV on the way out, sort of?"

After the crew entered the Lunar Module in a routine manner which would become standard practice during the Apollo missions, a television transmission was accomplished about 1 hour after ingress with a good TV picture – though not without its share of problems. As the television pass began, the image revealed the PLSS of Rusty Schweickart and the side of Jim McDivitt. Unfortunately for audiences the signal was subject to a large amount of interruption, yet when pictures were seen they were of remarkable quality and definition. The views afforded inside the Lunar Module were only ever seen during Apollo 9. All later missions would have the camera stowed outside the spacecraft and would only be deployed as the astronauts were climbing down the ladder onto the lunar surface. During this transmission, no voice was received at Mission Control due to incorrectly configured equipment at Merritt Island. The TV signal was unfortunately riddled with all sorts of communication problems, and several periods of picture disturbances. For TV audiences, however, the glimpses of the astronauts inside Spider were an enticing teaser for the upcoming to-be-televised spacewalk.

Houston advised the astronauts at 42:37:00 GET that they could begin the crew transfer to the lunar module; however, the crew replied that they were behind on the activities schedule. At 43:19:00, the Lunar Module Pilot was reported to be transferring to the lunar module. At 44:05:00, the Commander initiated the transfer. At that time, the crew was about 50 minutes behind in the flight plan, and all communications tests were cancelled except for the lunar module secondary S-band check and the two-way relay with television.

The plan called for the Lunar Module Pilot to egress, mount the 16-mm camera on the lunar module forward platform, transfer to and partially ingress the command module, retrieve thermal samples, transfer back to the lunar module, evaluate lighting aids during a dark side pass, obtain 70-mm still photography from the platform area, provide television transmission from the platform area, retrieve lunar module thermal samples, and ingress back into the LM. The entire operation was planned for 2 hours 15 minutes outside the spacecraft.

An artists rendering of the planned EVA. Note the 16mm camera attached to the CSM hatch. The TV camera would have been mounted on the LM near its hatch which is just out of sight in this picture (From NASA image S69-18547).

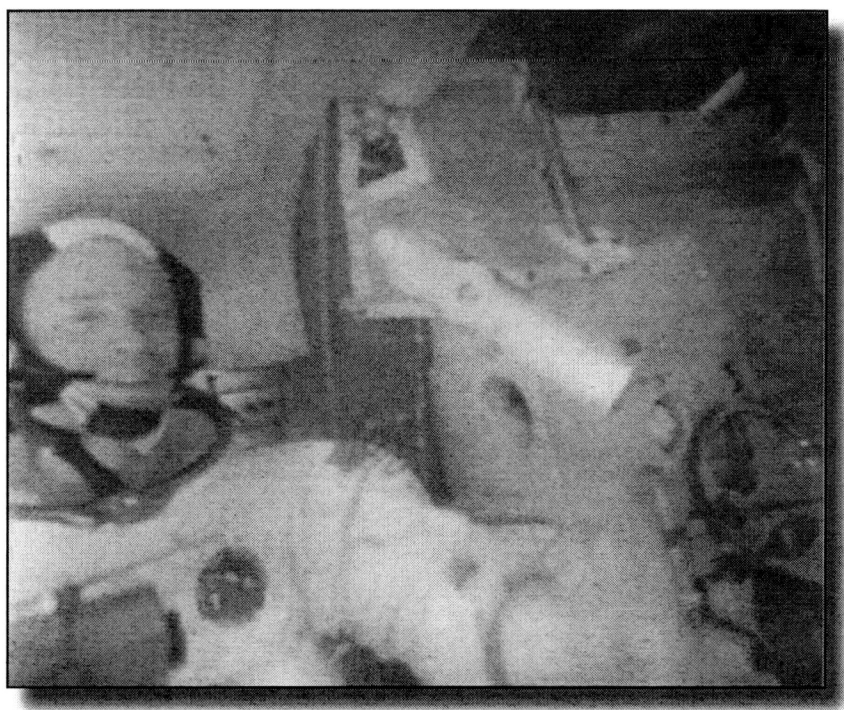

Slow scan monitor photograph of Rusty Schweickart with a radiation meter floating in the LM cabin during the TV transmission.

The EVA was cut short and was not televised as planned, although 16mm motion picture footage was taken by both Dave Scott and Rusty Schweickart during the egress. The second telecast began again with Schweickart and McDivitt standing in the Lunar Module during which the commander could clearly be seen eating his meal. During some general discussion with the crew, Mission Control advised that there would be a momentary loss of picture as the spacecraft passed through a region in the sky where no ground stations would be able to receive the picture.

When the picture returned with approximately 8 minutes left of useable signal reception, the camera was repositioned to show the back of the Lunar Module. Given the low light status in the back of the spacecraft the images showed remarkable clarity, just as Westinghouse had planned in such lighting conditions. To further demonstrate the diverse nature of the camera tube, the crew pointed the camera out of the Lunar Module window towards the Command Module. Orbital motion clearly evident, Dave Scott waved back to the two astronauts and an eager television audience back home. Loosely following the Gardiner memo recommendations, McDivitt attempted to show television pictures of the spacecraft; viewers had clear views of the quad thrusters, the Lander legs, and sundry exterior shots of both spacecraft from the window, along with the interior of the spacecraft. Sadly, the instrument panel shots were too dark to make out anything other than brightly lit panel lights, although the next shot would dazzle everybody watching: a side view of the command module silently flying over the earth.

Just as the crew was about to show the tunnel which connected both Spider and Gumdrop, the television signal was lost.

No further television was made during the remainder of the mission. Gumdrop and Spider were successfully undocked, and Spider was given the all-out tests necessary to qualify it for the upcoming lunar landing. All spacecraft systems operated adequately as planned. Aspects of the mission which included the firing of the ascent and descent stage of the Lunar Module performed well within specified restraints. The thermal response of both spacecraft remained within expected ranges for an earth orbital flight, which cleared the way from a problem-free flight to the moon and back. Management of the many complex systems of both spacecraft by the crew was very effective, and communications quality was for the most part very good. The TV pictures were clear and sharper than expected given the resolution and frame rate constraints of the system. As it transpired, the flight of Apollo 9 was the only pre-lunar landing test in space flight of the Westinghouse lunar TV camera. Luckily for television viewers around the world, and space historians for generations to come, a number of developments were to transpire which would take the next mission in a direction the engineers at Westinghouse quietly hoped would occur prior to the lunar landing. Television on Apollo 10 would be in color!

Television Downlink	GET hh:mm	GMT hh:mm	GMT Date
1st Transmission START	46:28:00	14:28	05 March 1969
1st Transmission END	~51:00	~14:32	
2nd Transmission START	74:58:00	18:58	06 March 1969
2nd Transmission END	75:13:00	19:13	

TV transmission times on the mission of Apollo 9. The end time of TV pass 1 is approximate and has been made using the actual transmission footage as a reference.

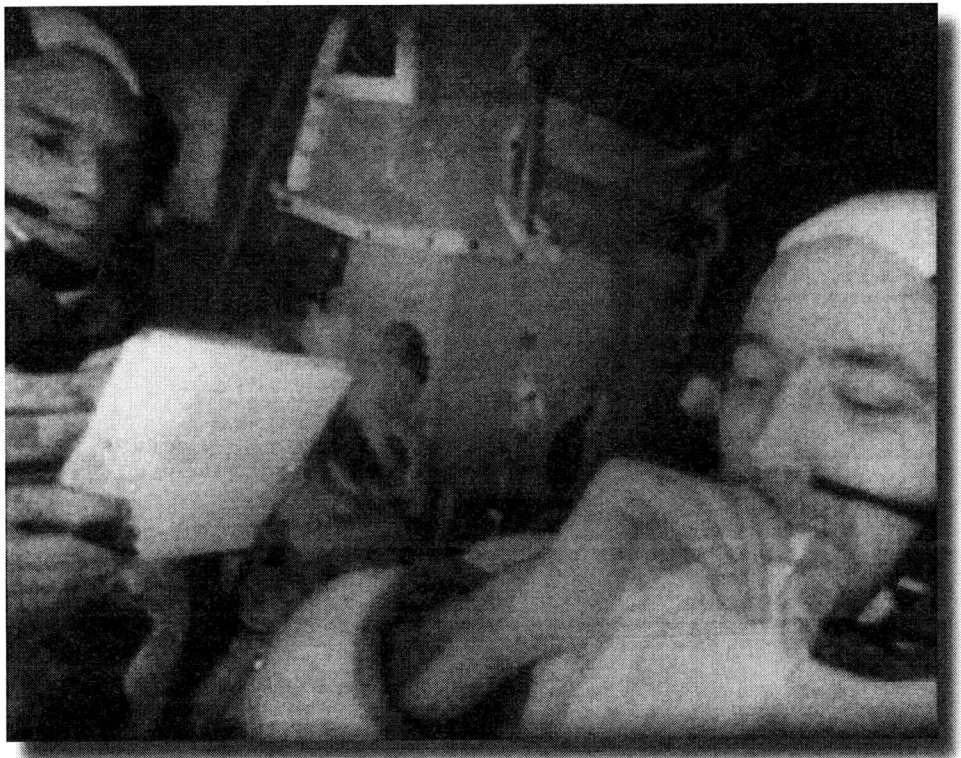

A frame from the 2nd TV pass on Apollo 9. Rusty Schweickart is on the left, while Jim McDivitt is eating some food on the right (From NASA image S69-26698).

CHAPTER 8. COLOR TELEVISION

"...we resurrected a TV system that had been mortally dead for some twenty years to solve a problem that could not be solved by the existing TV technology..."

Stan Lebar.

The idea of color television for in-flight television on the Apollo missions was not a new one. Of course, the limiting bandwidth parameters established very early in mission planning almost certainly curtailed any major research into camera development, however, it did not entirely rule out the enticing prospect altogether. Conducted over four months, "The Final Report for Color Television Study", prepared by the Astro Division of RCA, was submitted to NASA in March 1966. The report investigated several unique methods for sending back color television from the main locations during a lunar mission (earth orbit, lunar orbit and the lunar surface).

One particular method examined, known as the field-sequential color system, was a forgotten leftover from a television industry battle which occurred in the USA in the 1950's to secure a color television standard. An engineer by the name of Peter Goldmark, working for CBS with his team of researchers, had developed a color television system which used a spinning wheel in front of a single pickup tube in the camera. The wheel had three color filters of red, green and blue, which created a sequential order of color information. The receiving television sets used a similar spinning disc to create an image from separate hues as they arrived one after the other. In order to minimize a flicker effect brought about by each color being drawn separately on the TV screen an increased frame rate of 90 frames per second was introduced to seamlessly merge them into a full-color image.

The CBS system, by changing the frame rate, had effectively rendered all the black and white TV sets incompatible with the new color system. Additionally the new system would require a modified signal bandwidth to accommodate the new color information. Reinforced by huge sales in black-and-white television sets across the United States at the time, and fuelled by the incompatibility issues, the TV industry was reluctant to adopt it. Nevertheless, the body which regulated television and radio standards, the Federal Communications Commission (FCC) approved this system as the United States' first color television standard in October of 1950. Most engineers had concluded that the CBS system was below the acceptable standards of professional production, and were busy developing a more robust system, even after the FCC had approved it.

Competitor RCA, headed by David Sarnoff, wanted to dominate its share of the fledgling television market. Their vision had a much longer development phase for color, which allowed for complete backward compatibility with the black-and-white sets which were still selling at an enormous rate, while solving the color system riddle. Indeed, the manufacturers of these monochrome sets stood alongside RCA with their desire to have a compatible color system. RCA filed suit against the FCC in 1950, asserting that the CBS color system would be a backward step for the industry. However, the FCC was not convinced and in May 1951 upheld the previous ruling awarded to the CBS system.

It would be the Korean War which allowed RCA one last chance in its fight for a color system. Shortly prior to the outbreak of fighting in Korea, CBS acquired Hytron Radio and Electronics, in a last ditch attempt to gain the upper hand in television set manufacturing of its specific color wheel sets. However, with less than one hundred sets sold, the US Federal Government's National Production Authority declared the materials required for the manufacture of the television sets necessary for the war effort, and banned production of them. Using this decision to RCA's advantage, Sarnoff again appealed to the FCC and in 1953 won the ruling, which made his company's color television system the new U.S. standard.

The RCA system used three separate pickup tubes each being filtered as red, green and blue. The signal was then combined to form a full resolution color image with no artefacts brought about by a spinning filter-wheel, no need for field rate conversions, and no requirement for bandwidth adjustment. Most importantly it was backward compatible with the black and white television sets which had already been sold across the United States. The new system was a color version of the NTSC system already in use and would remain the standard for color television in the USA until the introduction of High Definition TV in the early 21st Century.

The CBS system would remain all but forgotten until project Apollo, when concerns about weight and power consumption on the spacecraft became paramount. As highlighted in the 1966 RCA Color Television report, the sequential alternative for reproducing color was thoroughly investigated. Other variants of color TV systems were also reviewed, such as the sequential information on each line of video information, by using both two and three camera tube systems and extrapolating color information electronically, and finally, a hybrid setup using the better aspects of the many systems detailed in the study.

Not surprisingly, the novel use of only one camera tube for color transmission made the field sequential system highly desirable, due to the obvious weight and power savings. A spinning color filter in the camera was used and the images were scanned in the sequence of red, green and then blue, to be subsequently joined together at the ground stations to form a full-color video signal compatible with broadcasting networks. However, a major drawback was the unavoidable color artefacts as a result of the filtered color signal. While research and design for a color converter were already underway, a working system had not yet been developed to place the separate colors into a holding pattern or buffer to form a complete image. Because such video disk storage devices were still the realm of engineering laboratories, the only advantage promoted at the time of the study was the potential for further investigation, once such devices were available on the market.

The study did however result in RCA developing an obscure variable line-sequential color television system, which used a three tube camera as the proposed method for obtaining color television. Using a frame rate of 7.5 frames per second, and 350 lines of vertical resolution, the required bandwidth was set at 1.25 MHz. It was concluded that this system would be ideal for the transmission of pictures in color via the limited bandwidth capacities of the Apollo spacecraft. This version of a color TV camera was submitted to NASA but was an impractical unit especially for use in the cramped Command Module. Citing the project as a non starter due to many design factors limiting its use in the spacecraft, the power hungry camera was rejected outright by NASA.

The report also deduced that any intended monochrome (black-and-white) signal should be obtained from the green filtered information, as it contained the most luminance information. It also found that a frame rate of 12 frames per second was found to be acceptable in all situations, while the lower rate of 8 frames per second was usually satisfactory in most TV scenes. Interesting to note is that initially the field-sequential system was overwhelmingly rejected due primarily to the color artefacts which were deemed unsuitable in all envisioned mission scenarios.

That, however, did not stop either RCA or Westinghouse from investigating their own methods for color television. Stan Lebar explains Westinghouse's research, "We knew nothing about the color system RCA had proposed until after we had demonstrated our sequential color camera. I was told that RCA had made an unsolicited proposal and apparently had crafted a non-operational model of the camera. It was a three tube NTSC camera which was described to me as very large, complex and a power eater. We discarded the concept of a three tube camera early on because of the issues as stated, as well as the optics system required to align the three images on the tube and remain aligned in a temperature swing of 500 degrees F."

For Lebar and Westinghouse, the desire had been to move away as quickly as possible from the severely limited black-and-white TV signal. Color television from the moon held a mystique which no black and white image could ever convey. Humans were going to be setting foot on another world for the first time in history, and for some this deserved a glorious celebration. Why should all aspects of Apollo enjoy state-of-the-art expertise, while TV images were relegated to highly limited technology? "I just had to find a way to get out of the Slow Scan B&W format and thought NASA could be interested in the use of color TV on Apollo instead..." explains Lebar. The single tube factor of a sequential color system was the key to keeping the weight and size of the camera within the constraints that NASA was extremely conscious about. Like the onboard computer which relied on the ground to carry out the resource-hungry calculations, Westinghouse moved the complex and bulky electronics required to create a color image from inside the camera to the ground stations. Quietly, while preparing the black-and-white TV camera for use on the upcoming Apollo 11 mission, the engineers at Westinghouse set about a way to garner NASA's interest in color television. Unbeknownst to them, they had someone on the inside who desperately wanted color TV from space.

Colonel Tom Stafford was television's best ally within the astronaut ranks. The immaculate beauty he had witnessed and commented on during his Gemini flights had impressed him greatly and he wanted very much to share his experiences with the tax payers who were funding the space program. In his book, "We Have Capture", Stafford explains, "What better way to take viewers along to the Moon than by using color television? But Deke [Slayton] and Wally [Schirra] had resisted the inclusion of a black-and-white camera on Apollo 7, though Deke eventually came to accept and even embrace the idea. Borman's crew carried the same fuzzy camera as Wally's did." He adds, "I thought we could do better."

In the months prior to Apollo 10, both Westinghouse and Tom Stafford were conducting their own schedule to have color television instated onto Apollo missions oblivious to the other's actions. Working hard with the development engineers, Lebar finally had a working color camera which employed the color-wheel technology borrowed from the CBS system and which used a newly developed disc recorder to act as a frame buffer for each color while the next in the sequence arrived. The colors were then added together via this buffering system which allowed the three independent red, green and blue fields to be shown simultaneously.

The color wheel mechanism found inside the Westinghouse sequential color camera. (Courtesy of Stan Lebar)

When the engineering team first conducted their shakedown test of the camera on February 17, 1969, they performed it in the clean-room which had previously been configured for the monochrome camera tests and as such, only had black, white and grey test patterns to use. Sitting in a corner in the room was an Eveready battery, which was then placed on a stand and used to evaluate the colors of the new camera. It was the only item around which had a variety of hues which could be used to test the spectral response of the camera. A Polaroid photograph of the monitor is all that remains of the moment. As Stan Lebar explains, "I have very special memories of that moment and although the Polaroid shot isn't that good, the actual colors that were displayed on the monitor was quite stunning and we were all taken by surprise at the color separation that we had been concerned about. We knew then that we had something special to demonstrate to NASA and we never looked back. The photo represents one of the more exciting moments for us all." All that remained was to contact NASA and try and sell the concept of a color television system which could be easily implemented into the mission plans of Apollo.

The first color images were captured on February 17, 1969, by the Westinghouse color television camera. The Eveready battery was the only thing in reaching distance which had a variety of color on it and so its bright blue and red colors were used for the inaugural test. There was no color converter used to render the image. (Photo courtesy of Stan Lebar.)

Lebar explains how the timing of Westinghouse seeking to demonstrate its system to NASA coincided with their search for a supplier with such a system, "My call to NASA (in early 1969) unknowingly to me, was made after the visit by RCA with their proposal and I was somewhat baffled by the immediate response. I don't believe the RCA proposal was ever shown to Stafford and to the best of my knowledge, the WEC sequential color camera was the first that he had seen that was an actual working color TV unit capable of meeting the small size, low weight and power as required by NASA and also capable of meeting the operational requirements for the Apollo spacecraft vehicles and could be upgraded to operate on the lunar surface. We demonstrated a complete system including the color converter so that he was able to see the color converted NTSC image that would be seen on earth as well as the unconverted image that he could see on the small mini monitor we had made just for his use in the CM. He operated the camera, zoom lens and mini-monitor and bought it without hesitation as long as we could make it meet the space environmental requirements as well as operate within the Command Module without interference bands in the image due to incompatibility with the vehicle electrical system. That was all thrashed out at KSC with the help of the manufacturers of the Command Module, the LM, the Saturn V and NASA. Tom Stafford brought everyone together and kept the program moving until it met its final testing in the CM on the stack prior to launch. The camera would never have been flown if it had been left to just those at NASA and had it not been for the efforts of Stafford to make it a reality. I guess our timing was just right but at the time of my call to NASA to let them know we had a working color TV system, I had no idea that it would end up being used on Apollo 10 nor did I have any knowledge that RCA had made a previous proposal earlier which had been discarded as impractical or know that Stafford had asked our counterparts at NASA to locate a color TV camera system that he could fly on Apollo 10."

An alternative sequential color system was demonstrated around this time by a company which had already unsuccessfully attempted to secure the contract for TV development years earlier in Apollo mission planning. Gyro Dynamics had an optical method for achieving a color signal and it was their demonstration which had started the impetus for NASA's switch to color television. However, the manner in which Gyro dynamics forced non-disclosure agreements upon NASA did not go down well, and given Westinghouse's track record with camera technology, the small firm was quickly dropped from consideration for a color TV system.

Stafford in the meantime had proposed his idea of televising color pictures from the Apollo spacecraft while on its journey to the moon to George Low at NASA. Luckily, the idea was met with a positive reaction, but under the condition that the color TV system be completely flight-checked and ready in time for the Apollo 10 mission in May, 1969. There would be no second chances and no excuses. If the color camera couldn't fly on the mission, the slow scan black-and-white camera would.

The engineering model of the color camera to fly on Apollo 10.
(NASA Photo S69-33917)

While the building of the camera would be simple enough--basically all that was needed was a synchronised wheel spinning in front of the pickup tube, the conversion equipment necessary had the research team at Westinghouse working overtime. The SEC tube was used, borrowed from the military development division and implemented into the camera to allow for low light use. The camera also met all the requirements for use in space and was extremely rugged and compact considering it was going to provide a color image.

One very important addition to the camera was the zoom lens. This would allow the astronauts to get a closer shot of whatever feature they chose, such as the Earth. The problem with such a lens being that it is nearly impossible to point the camera and frame the scene correctly without the astronauts having a monitor to view their actions. Integration of the zoom lens was simple enough, such technology was already in existence, but the monitor envisioned posed the greatest challenge for Westinghouse.

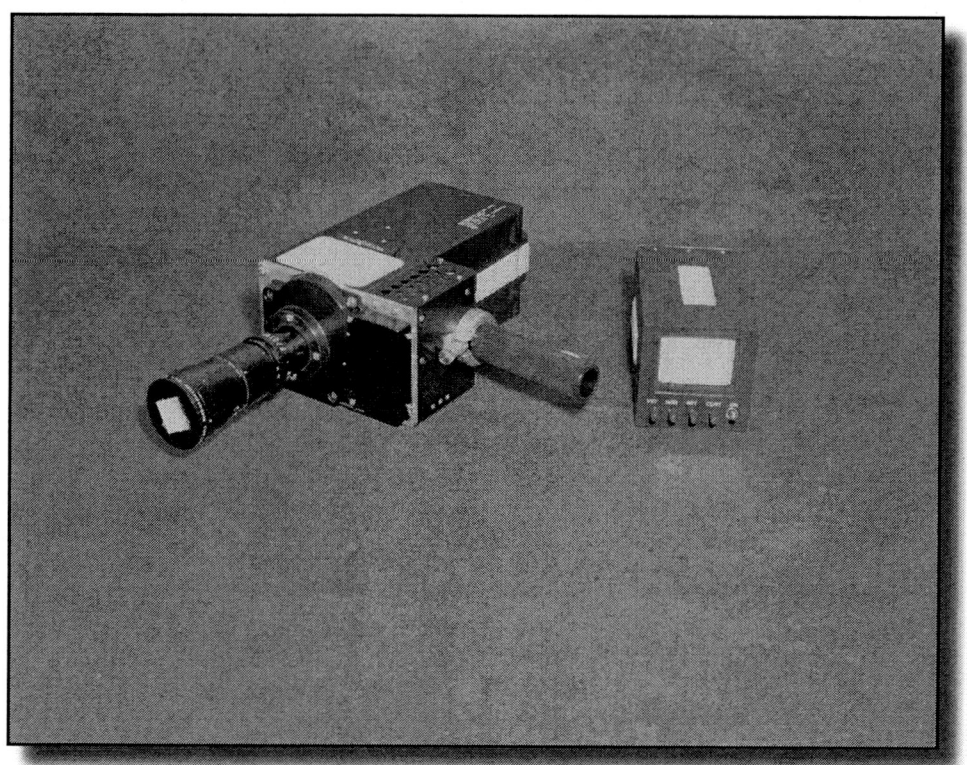

(NASA Photo S69-33916)

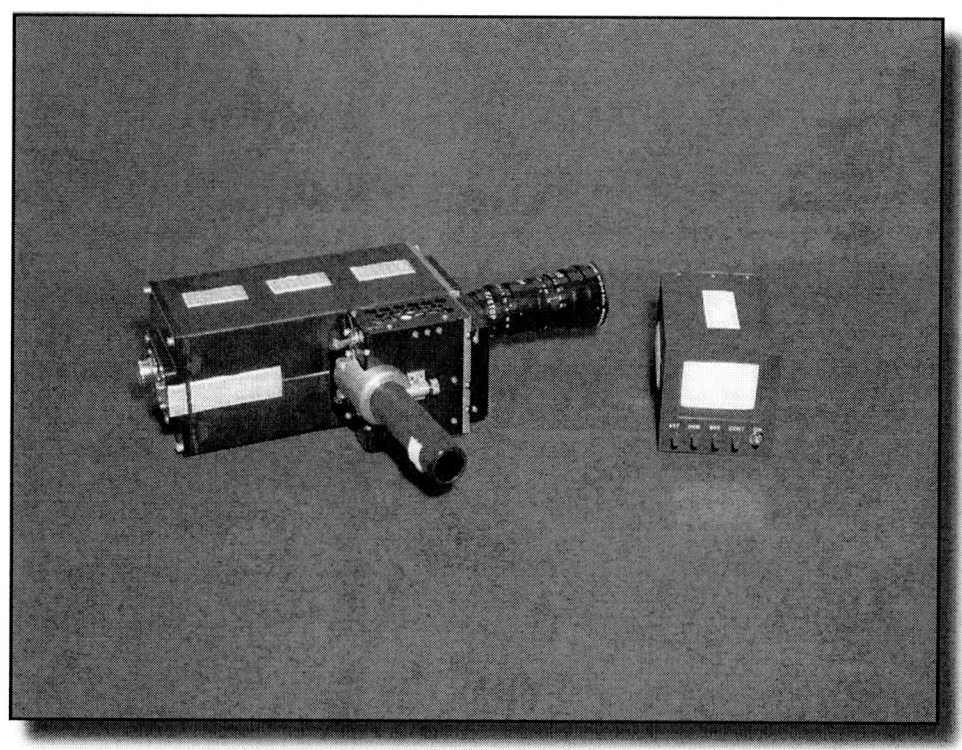

Apollo 10 Color Camera with Mini Monitor. (NASA Photo S69-33918)

Miniature monitors were very much relegated to engineering test benches. Investigation by Westinghouse uncovered an individual in Tokyo, Japan who crafted mini monochrome monitors. They quickly ordered several units and custom modified them into a device which resembled a tiny TV complete with brightness, contrast, and video controls much like a living room unit, making it very user friendly for the astronauts to operate. Initial pictures of the prototype show the monitor labelled as the "Runar Monitor" poking some stereotypical fun at the fact it was made in Japan.

The prototype "Runar" monitor (above) and the finished model (below) which was a little more serious, featuring a screen the size of a credit card protected by fire proof glass due to safety concerns inside the Command Module.

It could be attached to the camera to help in correctly framing the required shots. Gone were the experiences of Apollo 8 where directions for camera framing had to be given by the ground in Houston.

In an effort to demonstrate the color capability without the use of the converters to NASA representatives, Westinghouse decided to reveal its new camera with a specially modified monitor which could display the separate colors almost seamlessly and thereby create a faux-color image. A monitor was rigged to receive the incoming sequential colors and let the viewer's eye and brain combine them into a proper full-color image. To accommodate the starting sequence of the color wheel (when it was turned on it was never known on which color of the filter wheel would start) a switch was included on the monitor which could be changed to get the sequence to match that occurring in the camera. This could be selected to "RED, GREEN, BLUE", "GREEN, BLUE, RED" or "BLUE, RED, GREEN". No reliable electronic method existed to check the sequence and so engineer Phil Hoffman, who worked at Westinghouse Aerospace Division in Baltimore, Maryland, designed the switch to synchronize the incoming colors (unfortunately, despite Phil's schematic diagrams no official NASA corroboration has yet been located).

Lebar explains how it worked, "In effect the switch would actually either delay or advance the color wheel one color filter. Switching to the correct color phase took at most a second or two as it was so very apparent when you had the wrong color phase and when the correct color phase was selected...the monitor color image just seemed to burst into the correct color...there was never any confusion. This had to be done every time the camera was first turned on during the mission and it was accomplished so fast that it was never noticed or apparent."

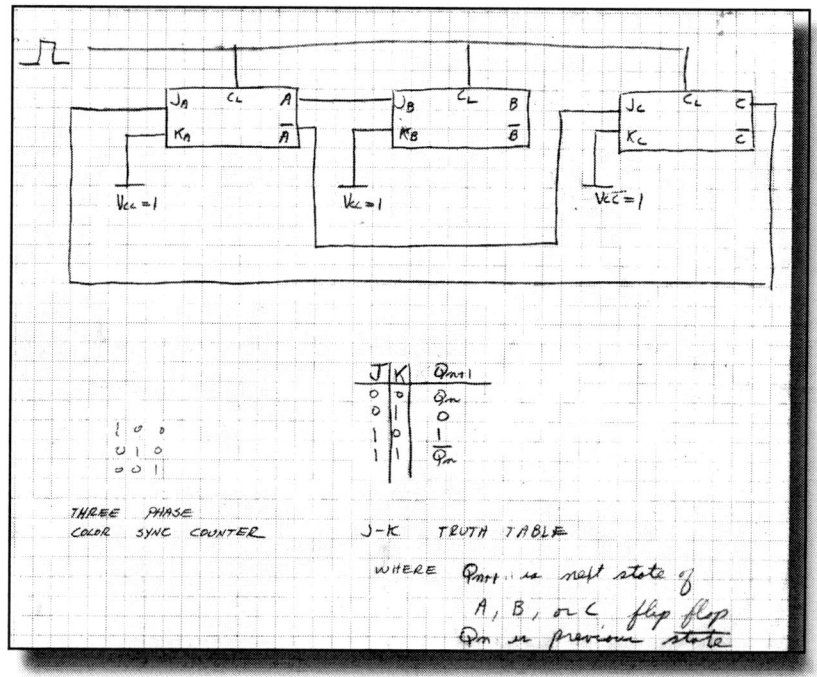

The switch circuit which allowed correct colors to be reproduced from incoming sequential Red, Green and Blue fields. Schematic diagram courtesy of Phil Hoffman

What remained was to build an operational color converter based on the video disc recorder technology which had been commercially available since 1967. Ampex had a system on the market known as the HS100 which could record 30 seconds of video information on a rotating disc, and play this information back at varied rates. It was designed to be used for slow-motion on sporting events, and through a clever adaptation of the technology involved in replaying the individual frames, NASA custom-built the necessary frame storage equipment, assisted in part by CBS Laboratories, to combine the red, green and blue sequential fields.

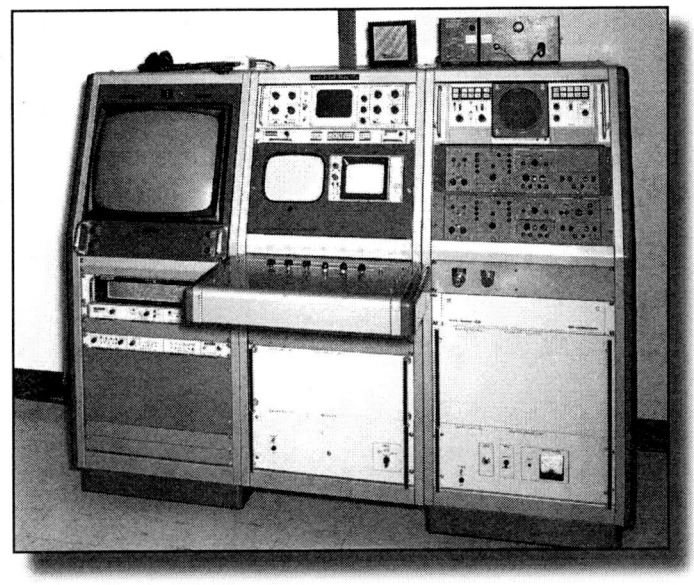

The Field Sequential Color Converter built for NASA.

(NASA photos S70-31960 and S70-31962)

A normal NTSC TV frame occurs 30 times per second which is made from two fields containing half of the image information occurring 60 times per second. As mentioned earlier, the first field contains every odd line (1, 3, 5 and so on) of information, while the second field contains every even line (2, 4, 6 and so on). The disc recorder recorded each incoming video field (1/2 of a complete frame) onto one of six available tracks. As field 1 was being written, 3 were being played back, and 1 was left blank. In order to recreate a full NTSC frame, there must be 3 even fields and 3 odd fields in the disc recorder. Due to the manner in which the fields were recorded in the first place, the field combination would always be wrong, that is the combination would always be either 2 odds and 1 even, or 2 evens and one odd. To rectify this, the incorrect field was always delayed by half a line, and thus it conformed to the other two fields.

The signal was then fed to an encoder made by Cohu which combined the fields of the respective red, green and blue fields to create a complete full frame of color video completely compatible with standard NTSC television sets. This signal could be sent on to any television station. In the case of non-NTSC format countries, the signal was converted one more time into either PAL or SECAM, although this occurred in the respective countries and was not performed by NASA. An additional problem had to be overcome as the spacecraft was travelling either to or from the earth. The Doppler-effect, whereby the frequency of the radio waves increase or decrease depending on the direction of travel of the transmitting spacecraft from the earth, could potentially cause timing problems to the incoming TV signal. In order to overcome this, a series of two video tape machines were linked together with one recoding the signal onto 2" tape and synchronized to the spacecraft signal. This tape was then wound onto the second machine synchronized to ground-based equipment and played back. The resultant signal was a fully compliant TV signal which would cause no problems in re-transmission. There was however a 12 second delay introduced for the color signal, which is often referred to in ground-to-spacecraft communications.

Disc Channel	1	2	3	4	5	6			
Field Designation	Ev	Od	Ev	Od	Ev	Od		Ev	= Even Field
Field Color	Re	Gr	Bu	Re	Gr	Bu		Od	= Odd Field
								E	= Erase
Disc revolution 1	E	R	R	R	O	W		R	= Read
Disc revolution 2	W	E	R	R	R	O		O	= Open
Disc revolution 3	O	W	E	R	R	R		W	= Write
Disc revolution 4	R	O	W	E	R	R		Bold	= Converted
Disc revolution 5	R	R	O	W	E	R		Re	= Red
Disc revolution 6	R	R	R	O	W	E		Gr	= Green
Disc revolution 7	E	R	R	R	O	W		Bu	= Blue

The read/write sequence pattern of the disk recorder used to create a color image from the sequence TV signal.

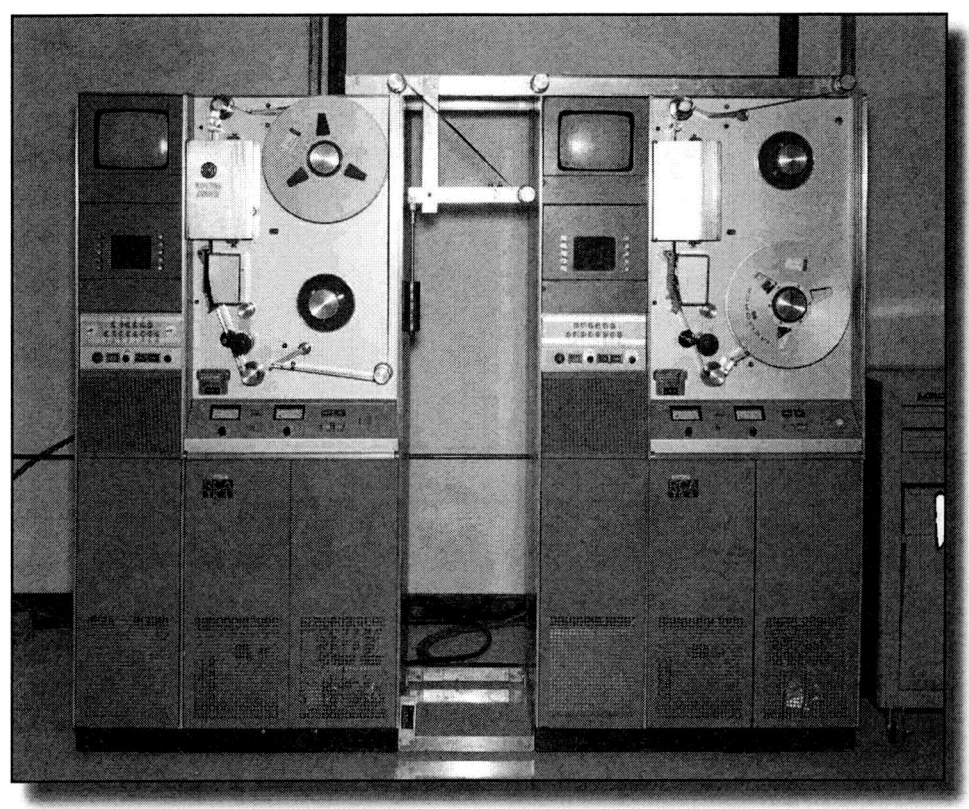

The dual 2" Doppler effect conversion setup. Note the pulley system to accommodate the increasing tape speed difference. The Machine on the left recorded the signal from the spacecraft, while the machine on the right replayed this signal to the ground-based equipment at NASA. (NASA Photo S70-31961)

Of course the real catalyst for getting the Westinghouse system approved occurred when Tom Stafford attended one of the demonstrations Westinghouse performed of the TV camera. Using the modified TV monitor wired to it he was shown color images from the camera which Westinghouse hoped would fly on a later mission. Their research team had placed their bets on the single tube sequential system as the only practical solution to color TV on Apollo missions. Tom Stafford was also aware of this and had become highly enthusiastic because of the demonstration he had witnessed. Directly due to his work in pushing to get the camera on his flight, NASA gave the green light for color television on Apollo 10 and suddenly Westinghouse found itself confronted with a two month time limit to have the system up and running for the flight.

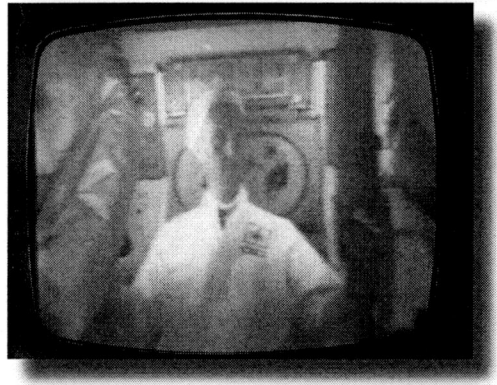

Color Television Test the Command Module LEB mock-up at the Kennedy Space Center.

(NASA Photos S69-32535 to 32538)

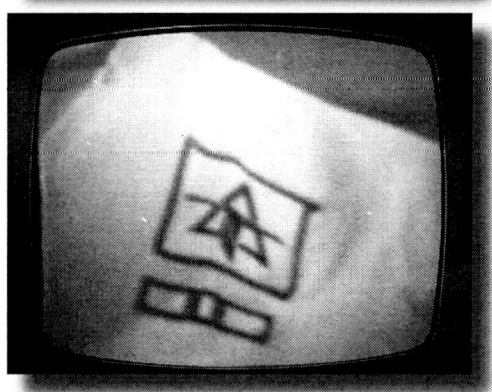

CHAPTER 9. APOLLO 10

"If all goes well after our countdown demonstration tests, we hope to carry a color camera aboard... which hopefully will give you a feeling...an actual visual interpretation of what we're seeing when we're up there."

Gene Cernan Pre Flight Debriefing 1969.

Tom Stafford had managed to motivate everyone at NASA to get the color TV included on the flight of Apollo 10. The blasé attitude now seemed to rapidly be disappearing into the woodworks as Westinghouse worked feverishly to flight-qualify the entire color system prior to the May launch window. Two separate internal memos by J. T. Raleigh and Robert Gardiner from April 7, 1969 were highly optimistic that the camera would be ready to fly. A response by George Low two days later indicated that while he strongly supported Raleigh's and Gardiner's views, the final decision for the new system's inclusion would not be made until a successful run of the Countdown Demonstration Test where the camera would be run through end-to-end tests. With Low's estimated investment of $200,000 of which $40,000 was set aside for the camera and monitor alone, the stakes were high for complete success during testing.

Gardiner in his April 7 memo had presented a timeline of expected delivery dates for the various pieces of equipment required for the flight of the TV camera. All camera equipment would be ready by April 18 for all aspects of formal testing at the Manned Spacecraft Center, with all ground support equipment all set on April 12--a good month prior to launch. All-out system tests were planned to begin April 21, where the communication lines used to transmit the TV signal could be examined. Additionally, flyover tests performed at Goldstone, using TV camera simulators (a test signal generator) would be made on April 28. Additionally, on the same day a test transmission was made from Goddard to Houston with the engineering model camera. An internal Westinghouse memo from Houston immediately following the reactions of NASA reads, "...the results were excellent...Most surprising to everyone was the lack of color breakup or streaking due to motion." The final major system analysis would be made on May 3, 1969. Stan Lebar recalls, "By the time the color camera was delivered to KSC, NASA had gained a great deal of confidence in the camera and the color converter due to the extensive testing that was performed at WEC as well as the testing that was performed at KSC early on and the subsequent final testing in the CM at the launch stack where the sequential video was transmitted to JSC by MILA, converted to NTSC at JSC and the NTSC video returned to JSC for review by all including the crew."

Westinghouse engineers had risen to the challenge, and through luck in the perfect timing of suggesting to NASA they had developed a color system which could be applied to existing camera equipment, they now found themselves supplying color TV for the next Apollo flight.

The NASA Manned Spacecraft Center Roundup newsletter confirmed the use of the camera on the flight by announcing just two days before launch, "Apollo 10 will transmit the first live color pictures of space through the major television networks next week. The new camera equipment was finally approved during the recent manned Countdown Demonstration Test and approximately 12 transmission opportunities were inserted into the flight plan. "We hope to be able to share with you some of the experiences and things that go behind the 'gee whiz' and 'golly, isn't that beautiful'," said Lunar Module Pilot Gene Cernan. "The camera is designed for use only in the Command Module and no LM television, color or black-and-white, is planned until Apollo 11. Because of the rapid development cycle of color, it is considered to be experimental in nature, and will be backed up by a black-and-white system."

Although the color camera did technically produce a monochrome signal, the output, when viewed without color conversion, had a very notable flickering effect. This discrepancy in luminance from one frame to the next occurred because of the varied light being scanned through the three different color filters. Of-course when combined, the differences merged to form a continuous brightness level as the different luminance levels from each color balanced out when combined. As the color wheel could not be stopped once the camera was powered up it was determined undesirable to use it as a monochrome camera. A memo by J.T. Raleigh strongly recommended the use of a black-and-white camera which, although having the reduced resolution and frame rate, would produce a flicker-free image in an emergency backup scenario.

Lebar refutes the notion of having the RCA black-and-white camera onboard by citing NASA's increased confidence in the color camera, "There was no WEC Slow Scan carried in the LM since it wasn't going to land on the moon and the only on-board camera in the CM was the sequential color camera. Since the switch from RCA Slow Scan to the WEC Slow Scan took place starting with Apollo 9, if they had need for a backup camera, they would have used the Apollo 10 WEC Slow Scan camera which would have been used if there had been no color camera."

Several tracking stations were planned for exclusive reception of the sequential color signal. Testing of the TV system concluded that a usable but very noisy picture could be received by an 85 foot antenna if the high gain transmitter was used on the Command Module. However, the use of the 210 foot antenna at Goldstone was highly desirable given the significant improvement in quality by using a larger receiver. Additionally Madrid was strongly considered for additional Apollo 10 coverage, and Honeysuckle was also available for use, provided sufficient planning and notice was given to the station.

On May 18, 1969 Apollo 10 launched with the crew of Tom Stafford, mission commander, John Young, Command Module Pilot, and Gene Cernan, Lunar Module pilot. Comprising of a highly experienced flight crew, the mission was to be the full dress rehearsal for the lunar landing mission with the sole exception that instead of landing, the Lunar Module would only come to within 11 nautical miles of the lunar surface. Just as promised, color television was used frequently throughout the mission, the first downlink occurring for Lunar Module extraction from the S-IVB stage. The first live colors from space showed the S-IVB booster still containing the Lunar Module.

The entire docking and extraction procedure was televised live to earth and performed flawlessly. A second downlink, nearly 30 minutes later, showed viewers more of the attached landing vehicle with a sharpness that allowed individual rivets to be seen on the spacecraft. It was the next transmission, however, which would leave many earth-bound viewers speechless.

Unlike the preceding missions which featured a low-resolution black-and-white circular disk, the television received from Apollo 10 showed the earth, our home planet, as a dazzling blue, brown and white ball floating against a sea of blackness.

Surface features such as Los Angeles, the Rocky Mountains and Baja were clearly visible in amazing detail. Capcom Charlie Duke could hardly contain his amazement as the images appeared on the screen. Audiences and news presenters around the world were glued to their screens. Nothing like this had ever been witnessed before on live television.

Although spoken years later in reference to similar images of the earth, not taken by the TV camera on Apollo 10, American Astronomer Carl Sagan summarized what this type of view of planet earth epitomised for humankind, "On it, everyone you ever heard of, every human being who ever lived, lived out their lives. The aggregate of all our joys and sufferings, thousands of confident religions, ideologies and economic doctrines, every hunter and forager, every hero and coward, every creator and destroyer of civilizations, every king and peasant, every young couple in love, every hopeful child, every mother and father, every inventor and explorer, every teacher of morals, every corrupt politician, every superstar, every supreme leader, every saint and sinner in the history of our species, lived there on a mote of dust, suspended in a sunbeam."

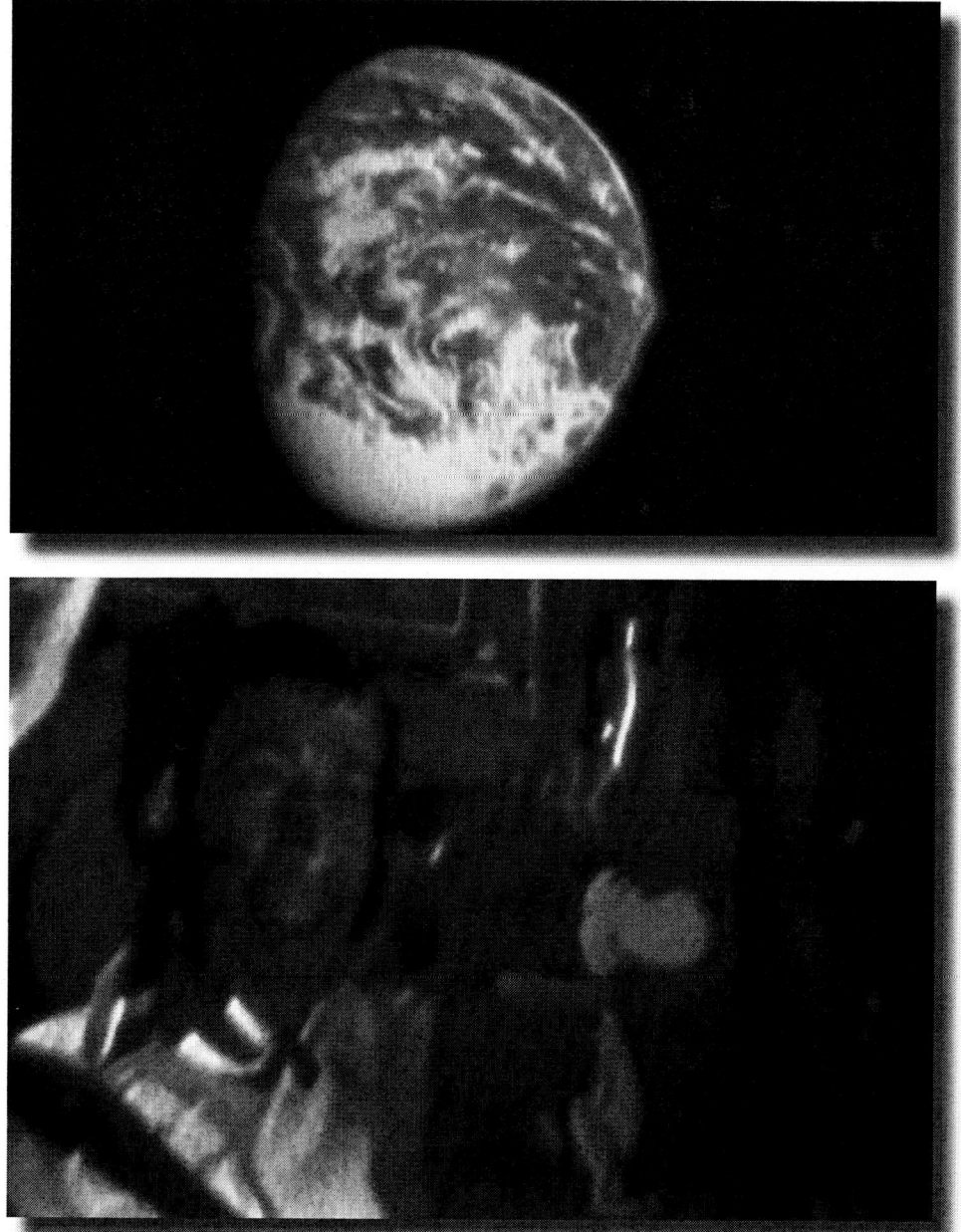

The images of a receding earth didn't stop after the first telecast. Six more televised views of our home planet showed an earth getting smaller and smaller as the spacecraft arrived from Apollo 10 on its way to the moon. When the camera was pointed inside the spacecraft viewers were treated to moments of levity by the astronauts. John Young held up a picture of the mascots for both the Lunar and Command Module: "Snoopy" and "Charlie Brown".

In one of the funniest moments of the combined telecasts, an upside-down John Young was bobbed up and down by a right-way-up Tom Stafford, to which he simply commented, "I just do whatever he says."

Experiments such as displaying air bubbles in a container bag filled with water were performed and general views inside the spacecraft were interspersed with earth views. The vivid colors were unmistakable as the Westinghouse camera delivered on its promise of outstanding colors from outer space each time it was turned on.

If that wasn't enough, once Apollo 10 attained lunar orbit, another transmission occurred which featured views of the lunar surface. Unlike the Apollo 8 TV images, the color view was striking. The lunar surface seemed to exhibit a brownish tint, which changed as the spacecraft approached the day/night terminator. Crater after crater was followed on camera by the crew with additional commentary provided by them to assist the observers on the ground.

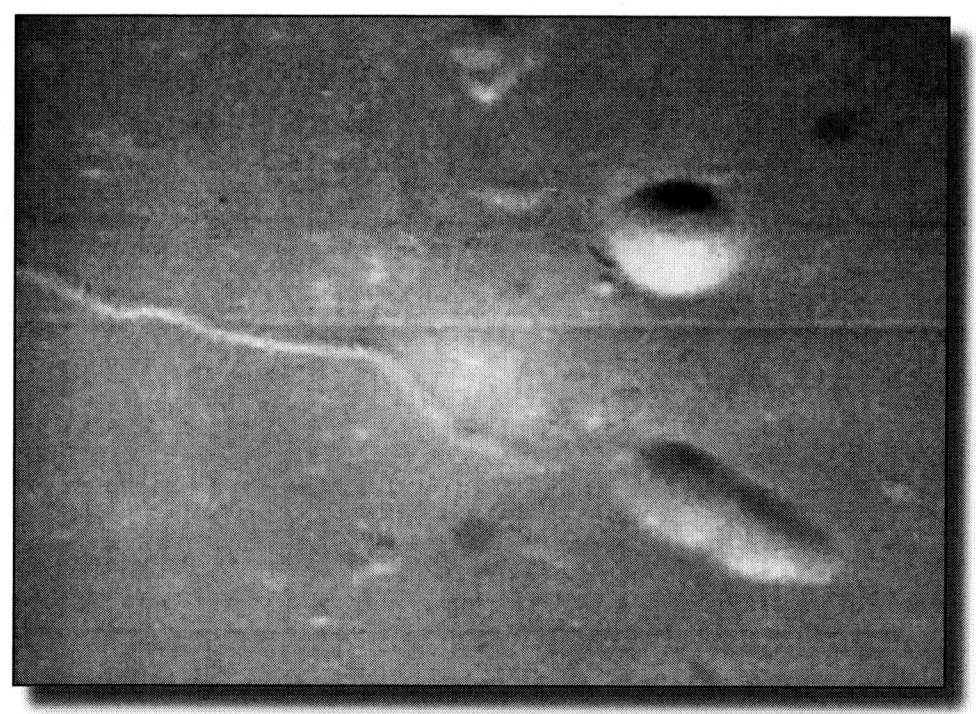

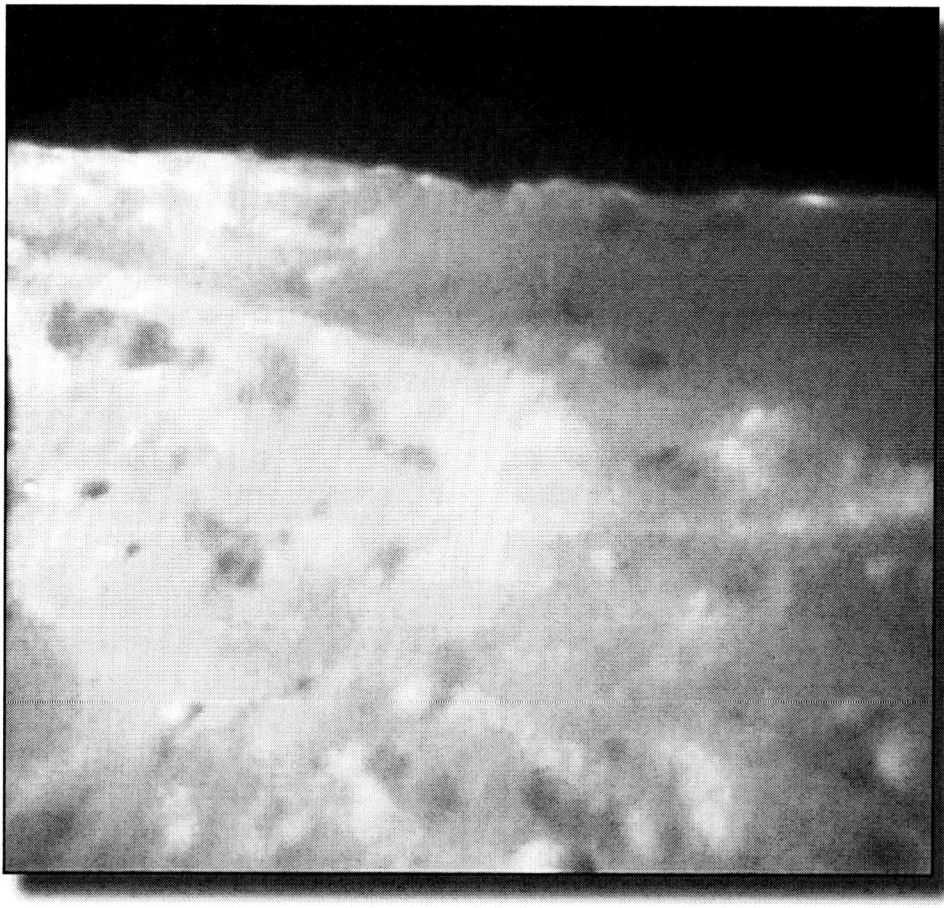

During TV coverage on the CBS network a pre-filmed interview with Cernan reinforced the images appearing on television screens all around the world, "I feel strongly, very, very strongly that this program is one which belongs to you, and your great grandmother, and your children, and the children of people sitting out there as well as to myself. I've been fortunate enough to be able to be a direct participant in it, but I like to be able to…share part of what's going on…hopefully to give other people a better understanding of what's going on and hopefully better to share some of that excitement that's involved. You can't help but be excited in seeing some of the things that are really happening right now and hopefully we can do this." Without question the addition of color made the whole mission seem so much closer and personal to those who watched the events unfold live.

Another transmission happened during the separation of the Lunar Module on its way to test flight above the lunar surface. A problem with the communication circuit between Goldstone and Houston resulted in difficulties in relaying crucial data for the spacecraft. An aggravated Tom Stafford, knowing full well the critical stakes in completing a successful mission bluntly told the technicians, via Charlie Duke to, "get with it!" Juxtaposed with the serene ballet of the LM which appeared upside-down, it was almost the realm of science fiction. The camera truly was offering a window into the regular daily procedures during spaceflight. Unfortunately, no television was supplied inside the LM, so voice signals were the only coverage to the most critical phase of the mission as Stafford and Cernan flew to within 11 miles of the surface, similar to the altitude of a commercial airline flying a daily route across the country.

An additional telecast featuring surface views of the moon occurred once Snoopy had rejoined Charlie Brown. There was significant discussion during the lunar scenes regarding the planned Sea of Tranquility landing site and the Apollo crew duly showed close-up imagery of the surface around the area. A further transmission after Trans Earth Injection showed a moon which was reduced in size compared to the previous TV show. While the pictures were being received on earth, a TV camera expert in Houston relayed information to the astronauts that the indoor/outdoor settings for brightness had worked perfectly. As the crew only had a tiny black-and-white monitor, they had no way of knowing just how brilliant their TV telecasts were. Of course, the numerous demonstrations on the ground had shown that vivid colors would be reproduced, it was only when they returned to earth, that the astronauts realised how much of an impact they had made on the general public.

Prior to their return however, the astronauts presented the world with further earth, moon and spacecraft views. The final transmission from Apollo 10 showed for one last time a large earth which filled the screen. While the planet filled the screen the crew once again reiterated how much it meant to them to share their experiences with the world. Houston relayed their thoughts on behalf of television audiences who now had a much better idea of what NASA was trying to achieve with their Apollo missions, and indeed that they were sad that the telecasts were coming to an end.

Looking back on the intense efforts to ensure color television flew on Apollo 10, Stan Lebar says, "I always regarded this effort as not only the most ambitious program we ever undertook on Apollo but the most cooperative TV program between WEC and NASA (not just JSC) that made it possible to go from a field sequential Engineering Model of the camera and Color Converter to being included as part of the Apollo 10 mission...all in just ten weeks. On top of that, it changed how NASA viewed Apollo television and changed the television format from Slow Scan B&W to NTSC color for all future manned space flights except for the Apollo 11 EVA which was never even a remote possibility of ever happening."

As the crew said their final farewells prior to their subsequently successful splashdown on May 26, 1969, anticipation was building for the next set of television signals to come from an Apollo spacecraft. The flight of Apollo 11 would essentially duplicate the TV transmissions seen on Apollo 10, with one major exception: a planned telecast as humankind first set foot upon the surface of the moon.

Television Downlink (View)	GET hh:mm:ss	GMT hh:mm:ss	Date GMT
TV Transmission 1 (LM docking) START	3:06:00	19:55:00	18 May 1969
TV Transmission 1 (LM docking) END	3:28:00	20:17:00	
TV Transmission 2 (LM extracted) START	3:56:00	20:45:00	18 May 1969
TV Transmission 2 (LM extracted) END	4:09:25	20:58:25	
TV Transmission 3 (Earth View) START	5:06:34	21:55:34	18 May 1969
TV Transmission 3 (Earth View) END	5:19:49	22:08:49	
TV Transmission 4 (Earth View) START	7:11:27	0:00:27	19 May 1969
TV Transmission 4 (Earth View) END	7:35:27	0:24:27	
TV Transmission 5 (Earth View) START	027:00:48	19:49:48	19 May 1969
TV Transmission 5 (Earth View) END	027:28:31	20:17:31	
TV Transmission 6 (Earth View) REC START*	048:00:51	16:49:51	20 May 1969
TV Transmission 6 (Earth View) REC END	048:15:30	17:04:30	
TV Transmission 7 (Earth View) REC START*	048:24:00	17:13:00	20 May 1969
TV Transmission 7 (Earth View) REC END	048:27:51	17:16:51	
TV Transmission 8 (Earth View) REC START*	049:54:00	18:43:00	20 May 1969
TV Transmission 8 (Earth View) REC END	049:58:49	18:47:49	
TV Transmission 9 (Earth View) REC START*	053:35:30	22:24:30	20 May 1969
TV Transmission 9 (Earth View) REC END	054:00:30	22:49:30	
TV Transmission 10 (Earth View) START	072:37:26	17:26:26	21 May 1969
TV Transmission 10 (Earth View) END	072:54:42	17:43:42	
TV Transmission 11 (Lunar Surface) START	080:45:00	1:33:40	22 May 1969
TV Transmission 11 (Lunar Surface) END	081:13:49	2:07:49	
TV Transmission 12 (LM Separation) START	098:29:20	19:18:20	22 May 1969
TV Transmission 12 (LM Separation) END	098:49:30	19:38:30	
TV Transmission 13 (Lunar Surface) START	132:07:12	4:56:12	24 May 1969
TV Transmission 13 (Lunar Surface) END	132:31:24	5:20:24	
TV Transmission 14 (Post TEI) START	137:50:51	10:39:51	24 May 1969
TV Transmission 14 (Post TEI) END	138:33:54	11:22:54	
TV Transmission 15 (Post TEI) START	139:30:16	12:19:16	24 May 1969
TV Transmission 15 (Post TEI) END	139:37:11	12:26:11	
TV Transmission 16 (Coming Home) START	147:23:00	20:12:00	24 May 1969
TV Transmission 16 (Coming Home) END	147:34:25	20:23:25	
TV Transmission 17 (Earth View) START	152:29:19	1:18:19	25 May 1969
TV Transmission 17 (Earth View) END	152:58:24	1:47:24	
TV Transmission 18 (Earth View) START	173:27:17	22:16:17	25 May 1969
TV Transmission 18 (Earth View) END	173:37:39	22:26:39	
TV Transmission 19 (Final TV) START	186:51:49	11:40:49	26 May 1969
TV Transmission 19 (Final TV) END	187:03:42	11:52:42	

The TV Transmission Schedule from Apollo 10. *The recorded segments were not seen live but recorded and played back at later time.

CHAPTER 10. APOLLO 11

"Oh thank you, television, for letting us watch this one!"

Wally Schirra co-anchoring the CBS telecast of Apollo 11 after Neil Armstrong set foot on the moon.

Apollo 11 was not only the culmination of nearly ten years of concentrated effort in realising a presidential challenge, it was also the attempt to fulfil a centuries old dream: to visit the moon. It rode on the shoulders of every spaceflight before it; every previous Apollo flight led up to the mission of Apollo 11. Every Gemini mission brought important manoeuvre and orbiting techniques upon which Apollo was designed. Every Mercury mission was necessary for the building of confidence in manned spaceflight. Unlike any other moment of exploration in the annals of human existence on planet earth, not only would the explorers conducting the mission be witness to the historical event, so too would the vast majority of humans still on planet earth watching images captured by a television camera on the lunar surface developed by Westinghouse.

From the day it was founded in 1886 by George Westinghouse, the company has been at the forefront of electric research and design. The United States government and Westinghouse had worked together prior to Apollo, mostly on imaging hardware which found its way into military applications in Vietnam. RCA originally had been contracted to supply the TV cameras which, it had hoped, would transmit the first television pictures from the lunar surface. Due to the close contact between U.S. military and Westinghouse, NASA had learned of advancements Westinghouse had made in video technology for low light surveillance.

The original contract for the development of television for use on the planned Apollo missions was awarded to RCA in 1962. Their camera was to be a small Vidicon tube camera operating at 10 frames-per-second and yielding a screen resolution of 320 lines. There was, however a major limiting factor which counted against the camera for use on the lunar surface. Potential use in low light scenarios effectively ruled out the RCA camera, as there was no way the Vidicon tube was sufficiently sensitive in such conditions. While Orthicon tubes were available at the time which would meet the requirements set by NASA for low light sensitivity, these tubes were in themselves still too large and bulky. In addition they were not built to withstand the shock and vibration which would be encountered during the launch phase of a lunar mission, nor were they able to operate in the lunar environment.

Westinghouse, in the meantime had already designed and manufactured video cameras which were specifically built to operate on the wings of supersonic aircraft, on the surface of a nuclear submarine, on navy vessels, and in the harsh battle environments encountered either on the ground, or on attack helicopters. The tube division of Westinghouse was also working directly with the Department of the Army in the development of the highly top-secret Secondary Electron Conduction Tube. At the Westinghouse Tube Division in Elmira, New York, Dr Gerhard Götze, a Research Physicist was in charge of building what would become the unique Apollo SEC tube. It was small, exceptionally sensitive in low-light conditions, and could operate automatically through the entire light level range. This made it an ideal candidate for NASA's lunar surface television requirements.

As early as 1955, Westinghouse had formed an Integrated Circuit facility with the goal of designing and manufacturing the complex integrated circuits the company had envisioned for military surveillance television technology expected in the 1960's. Designers of the television cameras took advantage of the facility and developed small IC designs into the cameras to be used by the military. These were lightweight, small, and required low power – ideal for military use which would require an extremely robust television camera for use in combat situations.

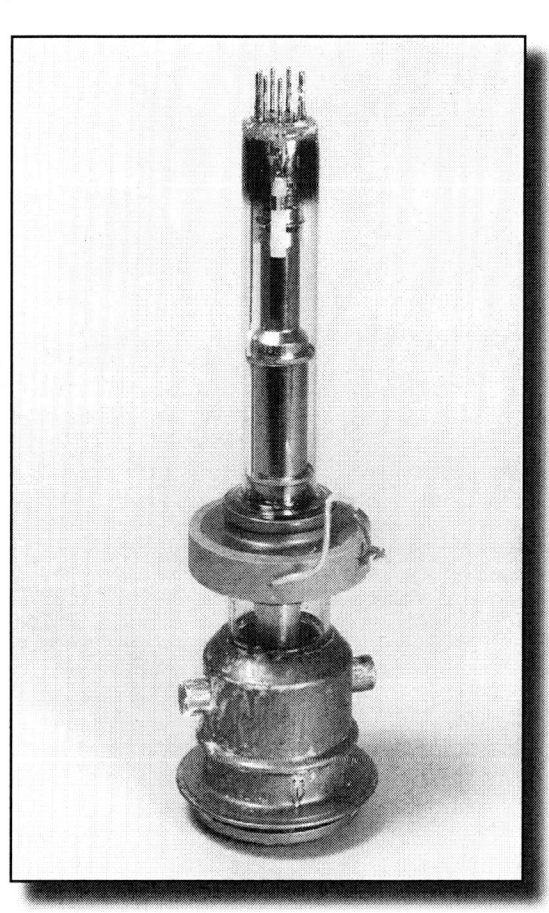

The Secondary Electron Conduction (SEC) tube manufactured by Westinghouse (Photo courtesy of Stan Lebar).

Reporting on the military applications of their television design, the Westinghouse Electro-optical group provided a dissertation on the camera developments at a television conference in 1964. The report highlighted the use of the Integrated Circuit facility to provide the necessary circuits for the specialized cameras, while also highlighting the advances made in television tube design. Upon hearing of Westinghouse's work, the company was asked to brief NASA officials at the Manned Spacecraft Center in order to ascertain whether the classified SEC tube could be adapted for the lunar camera television NASA hoped to operate on the lunar surface.

Shortly after the briefing for NASA, Westinghouse met with another contingent from NASA in Baltimore. Westinghouse was informed by NASA that they were the only company which could meet all the requirements stipulated for an operating television camera on the surface of the moon. These were:

- Small size and weight (6 pounds)

- Low power requirement (6 Watts)

- Low Light capability

- Thermal tolerance of temperature swings from -250 to +250 degrees Fahrenheit

- Automatic operation without the need for external control.

The contingency from NASA then asked if Westinghouse was prepared to provide the TV camera for use on the lunar surface, a question to which Westinghouse agreed.

NASA requested a proposal be submitted in 10 days from that meeting. In reality, in order to meet the deadline, Westinghouse had only seven days to complete the proposal. Stan Lebar was subsequently entrusted to oversee the proposal submission. He explains, "I had worked in the Electro-optical group prior to being transferred to the Space Department as a Program Manager and was the only Space Department Program Manager that was at the facility that week and I was asked to oversee the proposal. I was scheduled to make a final visit in two weeks to the Tracking Stations in Quito, Ecuador; Lima, Peru and Santiago, Chile to perform a final inspection of our equipment that had been installed at these sites. I was told that since the proposal had to be completed in a week, the effort would not interfere with my travel plans. When I had completed my visit to the Quito Tracking site, I received a call from WEC that MSC wanted to initiate contract negotiation for the Apollo Lunar TV cameras on Monday the following week, which was three days from that time. Since the planes to the U.S. were scheduled for only twice a week from Quito, I was advised to continue on to Lima, Peru and obtain a flight to Mexico City and connect to a flight to Houston. I did and flew overnight to Mexico City and thence on to Houston. The next day we started the negotiation at MSC, which was still being built, and a contract was agreed to three days later…"

Shortly afterwards, on November 2, 1964 NASA issued a news release announcing to the general public that Westinghouse had won the contract (allocated the number NAS 9-3548) to design the television camera for the planned moon landings.

> "HOUSTON, TEXAS -- The Westinghouse Electric Corporation, Aerospace Division, Baltimore, Md., has been awarded a $2.29 million contract for the development of a lunar TV camera."
>
> "This camera will utilize a recently developed secondary electron emission conductivity (SEC) Vidicon tube. It is considered to be the ideal image sensor for fulfilling the TV requirements during the translunar, lunar stay, and trans-earth phases of the Apollo mission. It is planned that these pictures would be made available to commercial television for nation-wide broadcast during the lunar mission, as well as provide scientific information."
>
> "Westinghouse was chosen as the sole source contractor after an intensive four-month review by a four-man committee from the Manned Spacecraft Center and the Lunar Excursion Module prime contractor, Grumman. All Government agencies and industrial firms who were known to be working in the area were contacted or their work was reviewed. The results of the committee study recommended Westinghouse because they are the inventor and developer and only manufacturer of a special image sensor which is the only sensor that can meet all the requirements of the lunar television camera specifications. These specifications include requirements in power, weight, and performance in the lunar environment."
>
> "For purposes of lunar operation, the camera must be designed to withstand temperature ranges from minus 300 to plus 250 degrees Fahrenheit. It must operate in high vacuum, and televise pictures in the glare of lunar day and the earthshine conditions of lunar night. The Westinghouse camera will be used in all Apollo flights in which the Lunar Excursion Module will be included in the spacecraft."
>
> "In Apollo Earth-orbital flights preceding those including the LEM, a camera provided by Radio Corporation of America under subcontract to North American Aviation, builders of the Command and Service Modules, will be utilized."

Wasting absolutely no time in the camera development phase, Westinghouse arranged a test on July 17, 1964, in which astronaut handling of the camera was assessed while they were fully suited. Hamilton Standard, the contractor hired to design the Apollo spacesuit, was to be on hand to facilitate the Westinghouse designers. Emphasis was place on arm and hand mobility of the pressure suit, no doubt to evaluate exactly how much practical use could be expected of the camera when operated by the astronauts whilst on the moon. The eye relief distance was also of great importance for designers to appraise the way in which the astronauts would be able to set up acceptable framing of a scene despite being physically unable to move the camera up to their eyes.

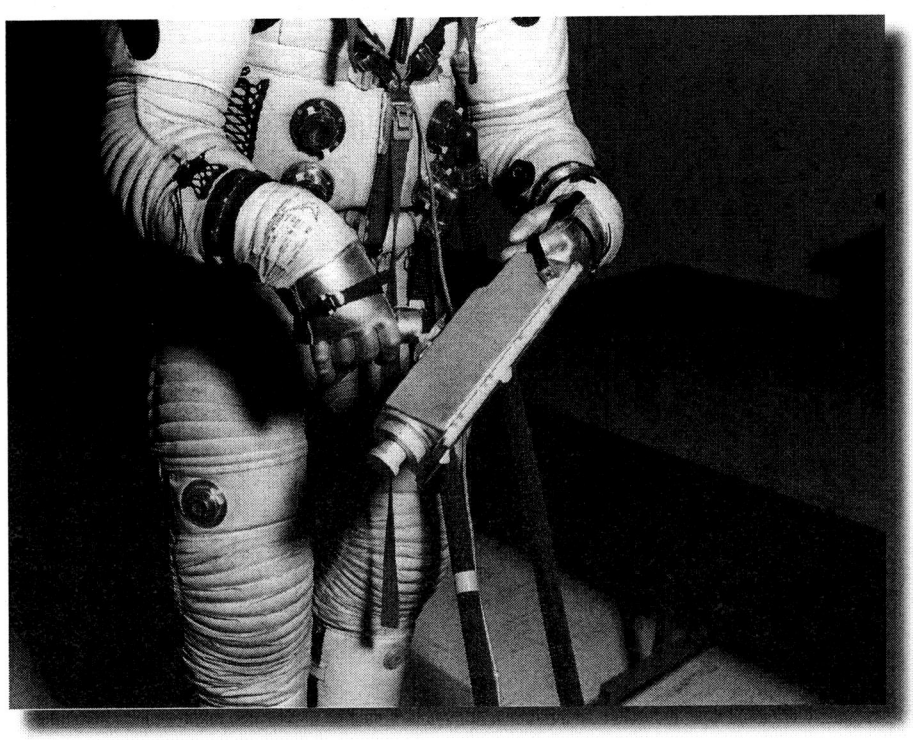

Tests done using a Gemini spacesuit to determine the design of the handle and power connector for the lunar surface TV camera (Photo courtesy of Stan Lebar).

In mid-December, 1964, a further operational proposition was made for the camera which would assist the evaluation of the lunar surface during the anticipated moonwalk. It was proposed that the Lunar TV camera be equipped with a high-resolution mode which would enable immediate pictures to be obtained containing greater clarity than the highly limited slow-scan TV image. Should the standard photo cameras have failed, this mode on the television camera would serve as an ideal backup system. A short period of discussion and demonstrations ensued before the green light was given to implement the feature on January 7, 1965. Adding an additional $200,000 to the TV camera bill, the high resolution mode was fully endorsed by Joe Shea and William Lee after clarification was given that the enhancement would place no burden on mission success. TV Subsystem modifications were also authorised for the Lunar Module in order to accommodate the possible use of the high-resolution function. The upgrade to the Lunar Module communications did not have any significant cost or schedule impact and so was quickly implemented.

Riding on their strong association with space cameras - particularly those used on the earlier unmanned Surveyor lunar missions, Bell and Howell submitted a proposal to NASA for their hoped participation in supplying the lenses which were planned on the forthcoming lunar cameras. It included over 100 pages of information pertaining to a wide angle, telephoto, lunar day and lunar night lens functions and the planned development of them. Similar to the television camera, the lenses were to be subjected to a number of environmental tests such as vibration, temperature and various lighting conditions.

Despite an impressive list of proposals the Bell and Howell lenses were ultimately not used by NASA.

The work to procure a suitable TV system for the lunar surface coverage was never-ending from the moment the contract had been awarded to Westinghouse. A deluge of design and research was undertaken to develop a camera which could withstand the expected vacuum of the lunar environment. By February, 1965, Westinghouse had a clear outline of the unit which would capture the first steps of an American astronaut on the lunar soil. Many of the operational constraints had been established and a preliminary design was featured which somewhat resembled the version of the camera eventually used on the moon. The camera was to have a trapezoidal shape in order to adequately keep thermal extremes under control.

The program, as outlined by Westinghouse, was to develop a television camera which would adequately meet all the Apollo Mission requirements placed upon it by NASA. The procedure proposed was to follow the sequence of developing a design, implementation of that design, followed, finally, by adequate testing which would successfully demonstrate the camera as being capable of operating on the moon's surface. Of extreme importance, was the close discussion with NASA's project office regarding any potential (and expected) changes which could occur as the planning of the missions became more concrete. Another benefit of this approach in communications was the ability to implement results obtained in the testing of other NASA programs should it have become necessary.

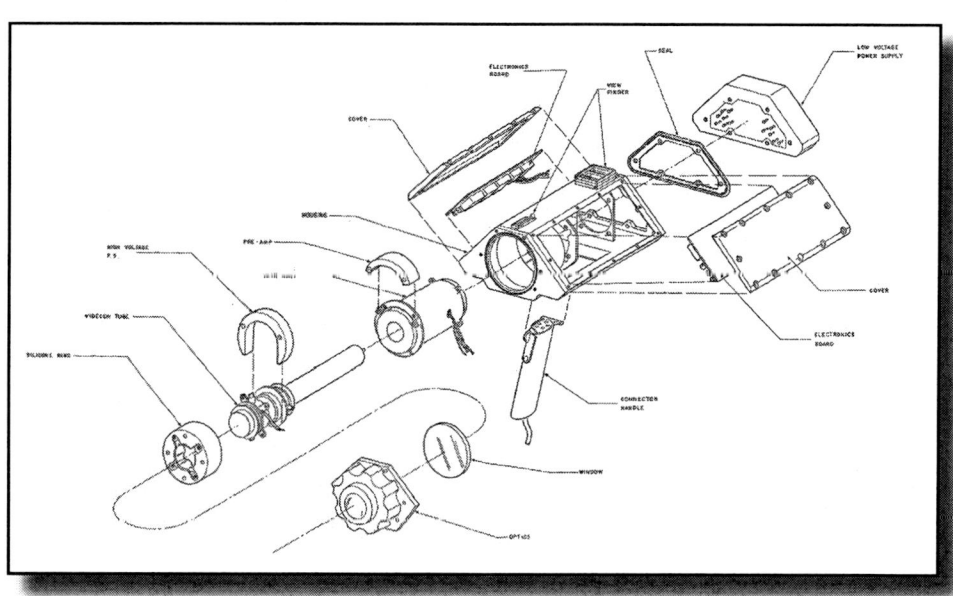

A cutaway image of the earliest known design of the Westinghouse Lunar Surface TV Camera. Note the inverted trapezoidal body-shape compared to the Apollo 9 & 11 TV cameras (From the WEC Lunar TV Camera Technical Summary No. 1, February 1, 1965).

Several aspects of camera development were apparent in the early reports conducted by Westinghouse. The electrical design was to concentrate on miniaturizing the circuits and implementing a fully automatic light level control operating in conjunction with the SEC tube. As extra functions were added, such as the addition of the selectable higher image resolution of 1280 lines was implemented, the size of the circuit boards naturally also increased. At the time of the report, however, such concerns were not major and the focus of electrical design was placed on the automatic light controls.

The mechanical design effort concentrated on the overall camera configuration. An increase in circuit size also directly affected camera design. However, the most critical importance was the passive thermal control which would allow uninterrupted camera operations in the temperature extremes the camera was anticipated to encounter during the lunar EVA. Studies were made into a variety of finishes for the camera housing which would facilitate nominal heat transfer for the camera.

Max Engert, NASA Manager of the Television System, holds a television camera mock-up, which is the forerunner of the equipment to be carried aboard the LM for broadcasting pictures back to earth. (NASA Photos S66-18729 and S-66-18730)

Westinghouse planned to use four fixed-focus lenses for the camera which could easily be interchanged depending on the lighting conditions. A night-time lens was to be used in lunar darkness, taking full advantage of the SEC tube and its high sensitivity to light. Three daytime lenses were also prepared: a wide angle lens for use in the cramped confines of the spacecraft, along with a medium focus and a telephoto lens for the lunar surface.

Unfortunately of 14 SEC tubes prepared for analysis, none of them met the stringent performance requirements. The resolution and light sensitivity factors plagued the tube design in the early phases of research, prompting changes to the overall tube design to address these issues. Nevertheless the preliminary study set the goalposts for the eventual camera Westinghouse would submit to NASA for the first lunar EVA.

It would not be the first lunar landing on which the lunar surface camera would be tested. The March 1969 flight of Apollo 9 was to fully test the camera in the airless environment of space. During the mission, it was hoped TV coverage would be made by Rusty Schweickart as he stood on the porch of the Lunar Module. Unfortunately the much anticipated telecast did not happen as planned, and only two short TV transmissions were made from the Apollo 9 spacecraft, and both from inside the Lunar Module. Despite a fairly elaborate sequence of shots to give a full shakedown of the cameras functions in the vacuum of space, only a brief glimpse inside the docked LM were afforded viewers on the ground.

(Photo courtesy of Stan Lebar)

Nevertheless, the camera had performed as hoped, and the notable light sensitivity improvement over the RCA camera was clearly observable in the transmitted images. The black-and-white Westinghouse camera itself skipped the flight of Apollo 10, thanks to Tom Stafford's insistence on telecasting color pictures for that mission. There was however a frenzy of testing on the ground to verify that the camera would be fully operational for the planned July landing of Apollo 11. Water tests, vibration tests, dusts and sealant tests all were ruthlessly applied to the small Westinghouse camera. Unlike most systems on the mission, the TV camera without any type of backup system would have to function properly the moment it was turned on. There would be no second chance

Every aspect of TV development into the Apollo spacecraft was carefully monitored by NASA. Seemingly mundane aspects of the TV system were meticulously debated and cross-examined until a suitable level of performance and spacecraft compatibility had been achieved. For example, the stowage location and connector interface requirements for the camera inside the spacecraft were given significant discussion throughout the early part of 1965. Of course, there could be no television from space if something as simple as the appropriate connections and cabling were not agreed upon. Discussion was continued until an April 22, 1965 memo directed the development of the cable, connecting the TV camera to the Lunar Module, be handled exclusively by Westinghouse. Grumman was in no uncertain terms to hand over material for the cable directly to Stan Lebar. The cabling was to be submitted first to MSC, who was then to give it to Westinghouse where it was to undergo full qualification tests. Additionally, the connecter interfacing to the Lunar Module was to be made by them and Grumman were to give the proposed costing to NASA, as they were no longer responsible for the item.

Mission planning was slowly being formed at this time as well. A Bellcomm report from May 6, 1965 detailed the planned use of the spacecraft communication system during the various anticipated flights once the project began flying missions in earnest. The use of television was intended as an integral part of the flight, albeit in a purely monitoring capacity. The report detailed two types of mission, an earth orbit mission and a lunar landing mission in which the entire communication subsystem was outlined in a cursory fashion.

Max Engert, NASA Manager of the Television System, further detailed the role Television was expected to play in an eventual Lunar Landing at the Institute of Electrical and Electronics Engineers (IEEE) Transactions on Aerospace in June of 1965. He outlined that, "The primary objective of the Apollo television system is to provide real-time television pictures of Lunar scenes which are suitable for viewing on commercial television receivers. The system would allow the general television viewing audience to follow the lunar exploration closely and to obtain a realistic idea of the lunar environment. The system will also supplement the historical documentation of the Apollo lunar mission. Other objectives include monitoring astronaut activities to obtain operational information for immediate and future missions. The initial lunar television system is not designed to obtain highly accurate scientific information such as close details of lunar topographic structure; however, some scientific information of a gross nature should be obtainable. It is apparent from the above objectives that a television signal of the quality usually seen on the average home television receiver is required, or at least desirable."

The preliminary plans, for televising the lunar mission were detailed W.E. Zubrek on July 31, 1965 at the NASA 1965 Summer Conference on Lunar Exploration and Science. These ideas varied somewhat for those which were eventually used on Apollo 11. For example, the use of television during the pre-launch and launch phase of the mission was to be performed. Some of the earliest recommendations at the start of mission profiling very early on had routinely mentioned launch television as being desirable. As detailed by Zubrek, there was to be only one slow scan television camera which would be transferred into the lunar module for the flight to the lunar surface. Once the astronauts had landed on the moon, the television camera was not to be activated until it had been removed from inside the Lunar Module, and an erectable S-band antenna had been set up on the lunar surface. Such recommendations did save crucial weight in the launch vehicle. Using only one camera certainly made stowage problems easier. However, the all-important first step onto the moon would not be seen live in this proposed configuration. Needless to say the planned operations on the moon were consistently under revision.

Apart from the camera tube, the most important component of the lunar surface TV camera would be the lenses, and August of 1965 saw a large amount of testing performed on them to evaluate their robustness on the lunar surface. Fairchild Space and Defence Systems had been contracted to conduct a series of tests on the lunar camera lens components. The company was required to fully qualify the components under stress conditions, such as launch and unusual environments such as the vacuum on the surface of the moon.

An inside view of the electronic circuits used in the lunar surface TV camera. (Photo courtesy of Stan Lebar)

(Photo courtesy of Stan Lebar)

By subjecting the lenses to a series of noise pressure variants, they could be analysed for resilience against extreme vibrations which were anticipated during the launch of the Saturn V rocket. Shock tests were also conducted again to simulate the effects of a launch. For corrosion qualification a series of salt spray tests were undertaken. Examination was then made of the surface of the lens components to ensure they had not deteriorated below the target levels outlined by Westinghouse. Minor details such as the focal ring flange on the lens were checked, along with resolution and depth of field of images made with different lens settings.

One crucial set of tests was to check the ability of the camera and its associated lenses to obtain good image quality in a variety of temperatures. These temperatures ranged from extremely hot, through to normal, and subsequently extremely cold. The resolution mode of the TV camera was also checked, with the temperature ranges applied to both the low resolution and high resolution settings of the camera.

The sealants for the lenses were rigorously subjected to integrity tests under the temperature variations. Further sealant examination was performed by Ardel in late 1965. The company evaluated O-ring seals for use in a vacuum environment. A report was submitted which tested different types of rubber seals in various gas atmospheres. Once the qualification tests were performed, a further set of examinations were made exceeding qualification target levels of tolerance.

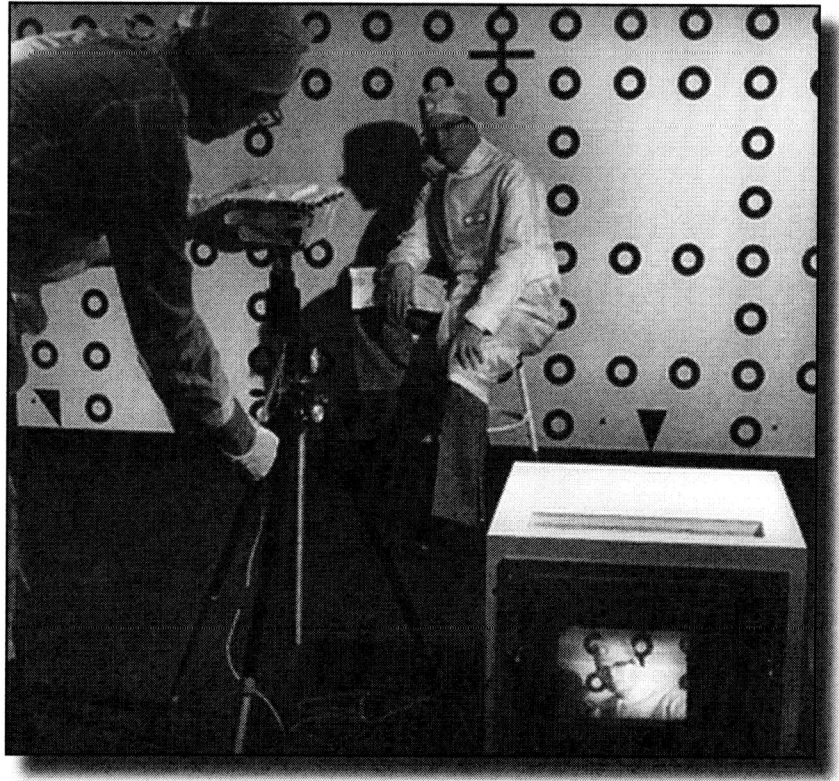

(Photo courtesy of Stan Lebar)

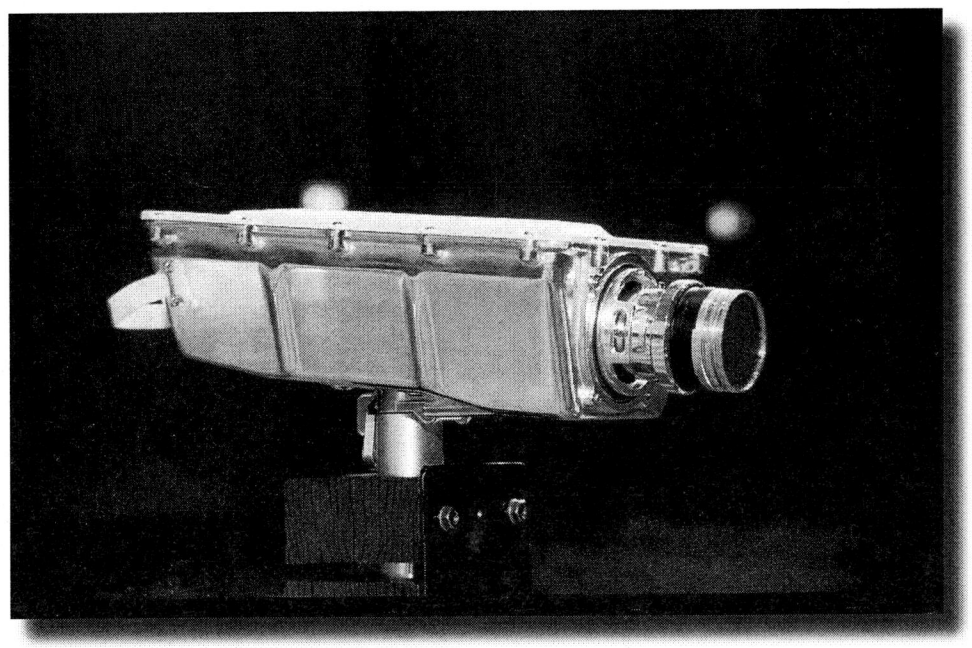

A close-up view of the small portable television camera which was used on the lunar surface to broadcast images back to earth. (NASA Photo S66-32993)

The findings of these test configurations were then sent to Westinghouse for review (and action if so required), and were summarized at the end of each month in the Westinghouse Progress Reports. Like all the other contractors working towards the goal of a successful lunar landing, the Westinghouse progress reports were a concise evaluation of the development of the camera. Problems concerning, say, tube performance, or pickup plate sensitivity were outlined and suggestions were made for action to resolve the particular dilemma. Factors such as the adherence of the finish to be applied to the camera housing were amongst the items covered each month.

On June 29, 1966 Max Engert gave a demonstration of the TV camera and explained the scan converter operations to interested guests at a meeting of the Instrument Society of America. Similar to his dissertation at the IEEE convention, he explained the logistics behind the evolution of the lower resolution and lower frame rate modes of the TV camera. It would not be until two months later, in August, that Westinghouse would have a full official report ready for NASA regarding the lunar surface camera. The Statement of Work report detailed the planned uses for the camera, and the current status of the device in design and development. The strategy in mid 1966 was still to have one camera servicing both the Command Module and the Lunar Module. The camera, even at the time of the report was to be mounted in the CM for pre-launch, although the requirement for a televised cabin during launch had been scrapped. Various phases during the journey to the moon were to be monitored, as had been the plan since the early days of devising the missions. The camera would then be transferred to the Lunar Module and taken to the surface. It would be activated once the astronauts were outside and then taken back to the Command Module once the mission had been completed and the Lunar Module had docked. The TV camera was supposed to work in either lunar night or lunar day, and it was assumed by Westinghouse that knowledge of whether the mission would be in either the day or night would be made early enough to allow small modifications to be made which would save weight, power consumption or equipment complexity.

As early as January, 1966, plans were underway to test a prototype TV camera on unmanned Apollo flights. John Noble Wilford reported in the New York Times, that NASA had cleared the camera for the flights planned for later that year. As Engert had demonstrated in June, the camera was going to require conversion from its 320 line, 10 frame-per-second scan rate to the commercial NTSC standard. Several companies had submitted proposals for a real-time converter; Westinghouse amongst them. One of the earliest known records of scan converter technology development was with Image Instruments Inc. of Massachusetts. They were awarded a contract by NASA in 1962 to study the conversion of slow scan television into a standard television signal able to be viewed on a standard living room television set. Lester C. Smith, president of Image Instruments announced that, "NASA will use this scan converter for evaluating the visual effect of displaying at standard rates those incoming signals which have been taken at a variety of different line and frame rates. By applying three cathode ray storage tubes in a time sharing fashion we will eliminate flicker and smear in the display. Incoming signals will vary from 5 frames per second to 25 frames per second and 100 lines per frame to 525 lines per frame." The purpose of the converter was to assist NASA engineers in establishing the best picture quality and selecting the most suitable compromise in bandwidth, equipment size and weight for future space television systems.

Westinghouse had offered an unsolicited proposal for a tube scan converter in February of 1964. Based on a hypothetical system incorporating components that were not yet developed, the system used the long storage SEC tubes to convert the lower frame rate image into a compliant commercial television system. It promised to be free of flicker, shading and smearing problems commonly associated with scan conversion. A six-month window of development time was suggested for the Westinghouse converter.

In the meantime, Lear Siegler, which had provided the slow scan TV camera for the one-time Mercury television tests, submitted their proposal for a fully self-contained monitor for viewing and photographing slow scan TV signals. One huge benefit of their unit was that it was to be calibrated at the factory, thereby requiring no major setup upon delivery. Minor adjustments could, however, be made on the unit via controls found on the front panel. These included settings similar to a home television set such as brightness, contrast, vertical and horizontal hold adjustments and minor correction switches required for the varying signals it was designed to handle.

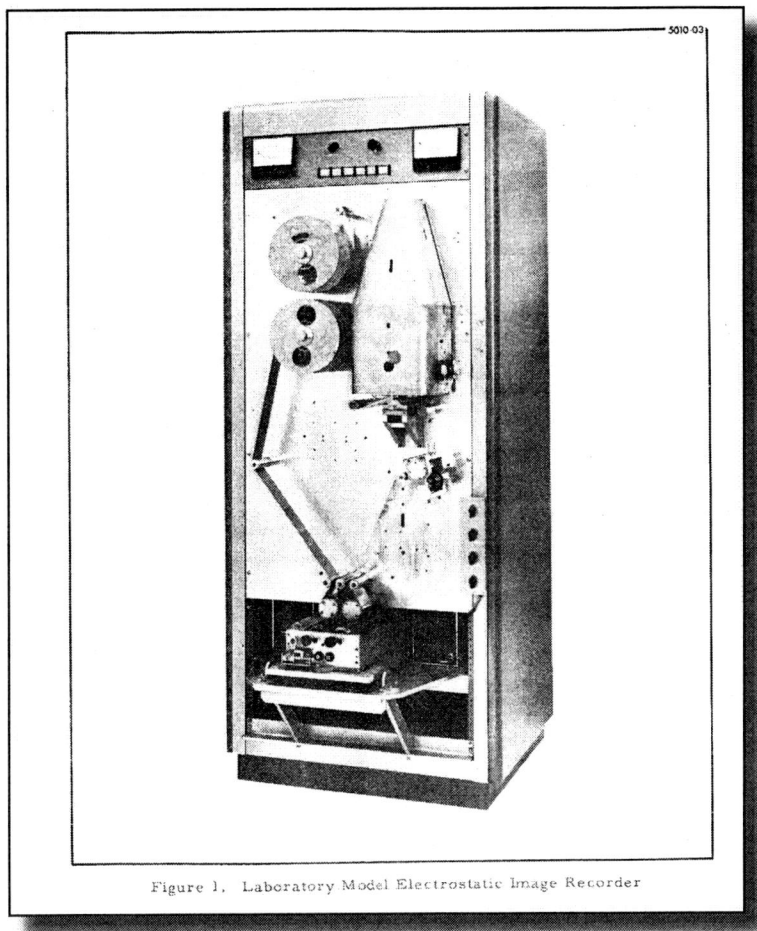

Figure 1. Laboratory Model Electrostatic Image Recorder

Figure 2. Scan Converter Demonstration System

Views of the electrostatic scan converter supplied to NASA by the Litton Company.

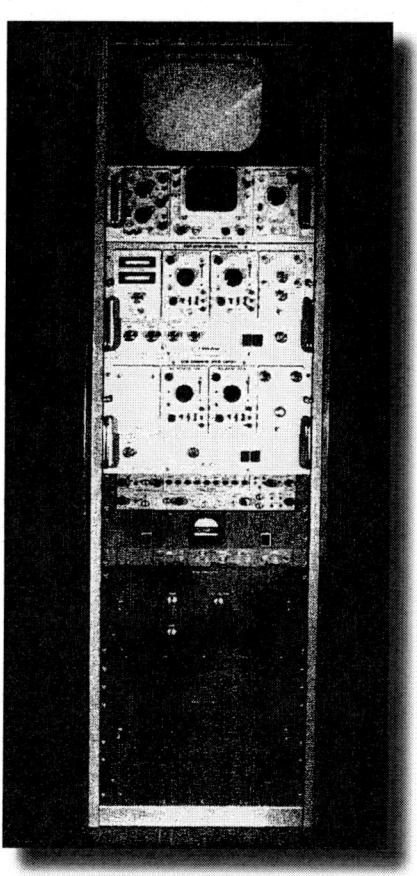

An extremely rare photograph of the prototype Westinghouse scan converter.

The Litton Company proposed an Electrostatic Image Recorder for television scan conversion in February of 1965. Their method of conversion would record the incoming 10-frame-per-second image patterns on transparent film, develop the image by applying a pigmented, electrostatically charged powder; and subsequently project this image into a television camera which output a standard NTSC TV signal. While the demonstrations of the device were most certainly promising, the picture unfortunately degraded to an unacceptable level. Litton felt the picture tubes were at fault, and aimed to resolve the degraded picture problem in its ongoing design of a prototype model for further demonstration. Despite the promising outcomes, NASA decided not to adopt the electrostatic scan converter.

By March 31, 1966 Westinghouse had a working prototype of their intended scan converter prepared for NASA. The model presented was indicative of the converter ultimately used on Apollo 11. It used a gated camera pointed at a high resolution cathode ray monitor, which then sent one field for each frame of the slow scan image appearing on the monitor to a storage device. This storage unit was a modified videotape recorder which then replayed the field continuously until the next field was written to it. The end result was a commercial video sequence made of one direct field of the video camera with the replayed redundant field from the video recorder. During the course of the design, Machtronics, who had earlier submitted a design for a portable tape recorder to be used on the Apollo spacecraft, designed a disc storage unit. It was used in conjunction with the videotape machine during demonstrations and testing. While leaps ahead of anything yet proposed to NASA, there were still image degradation problems which needed to be addressed prior to the unit being accepted as the scan converter to be used for the actual space flights. Nevertheless NASA was quite enthusiastic about its potential, submitting the unit to intense testing to verify it was fully operational. The cost of the Westinghouse converter was estimated at $98,000.

Fairchild, in addition to testing the TV lenses, also had developed their own version of a slow scan monitor in June, 1966. The monitor differed greatly from the other proposals in that it was only designed for conversion to a photographic medium: Polaroid or 35mm. It was more of a monitor for making immediate photography rather than providing a continuous stream of video information. The advantage of the system was its brightness, which permitted fast photography of the slow scan frame in question. It also permitted sequential frames to be automatically exposed onto one frame of film. What is unclear in the proposal, however, is whether the converter could be upgraded to allow real-time video conversion as well. All that is suggested in the report is that Grumman worked closely with its development and implementation.

NASA, at this time also pursued the testing of slow scan television information over commercial long lines. Photographs of various test images were provided and published in a report by M. J. Quinn in 1967. The images were evaluated for their quality and helped NASA to study how robust slow scan information was, when sent over long distances.

Despite the efforts of Westinghouse, their idea for a converter was ultimately never used during any Apollo mission. A scan converter built by RCA was the unit chosen to convert the slow scan TV signal from Apollo 7, Apollo 8, Apollo 9 and the Apollo 11 moonwalk. Their unit very similar to that devised by Westinghouse, used a stock standard video camera which had seen use in film-to-video telecine, and in the days prior to videotape was also used to record video onto film (a process known as kinescope). It was a black-and-white Vidicon tube camera pointed at a 10" high resolution cathode ray monitor. The monitor had a persistent phosphor which caused the image to remain on the screen for longer than normal. The TK-22 was gated to record 1 frame as it was written onto the high resolution screen.

The output from the camera was a standard interlaced NTSC video signal. 1 full frame of video information was composed from two fields of 262.5 lines which the camera could not properly record from the 10 frame-per-second rate. The first field was recorded correctly, but the second field would be recording off the monitor when the next frame of video information was already being written, resulting in a messy signal which generated a lot of problems in the conversion process.

This snag was overcome by recording the first field onto a video disc recorder which would then repeat the redundant field with a delay built into every second field to allow it to mimic the missing field that the camera was unable to capture. Essentially, the TK-22 recorded the first field, with the disc recorder repeating the fields while adjusting them so that they correctly formed a full NTSC image. This process was repeated to form the "missing" 3 frames of NTSC video and the resulting output was a fully compatible NTSC video signal. There was one major drawback, which unfortunately the technology of the time could not solve. The picture was unavoidably degraded as it was optically converted and this on top of the already reduced resolution of the incoming slow scan TV signal.

The system controls were rather straightforward. A test pattern generator was incorporated into the converter to assist in calibration prior to receiving an incoming signal. The type of slow scan signal could also be selected between the 10 and .625 frames-per-second rates, though in the case of Apollo 11, the high

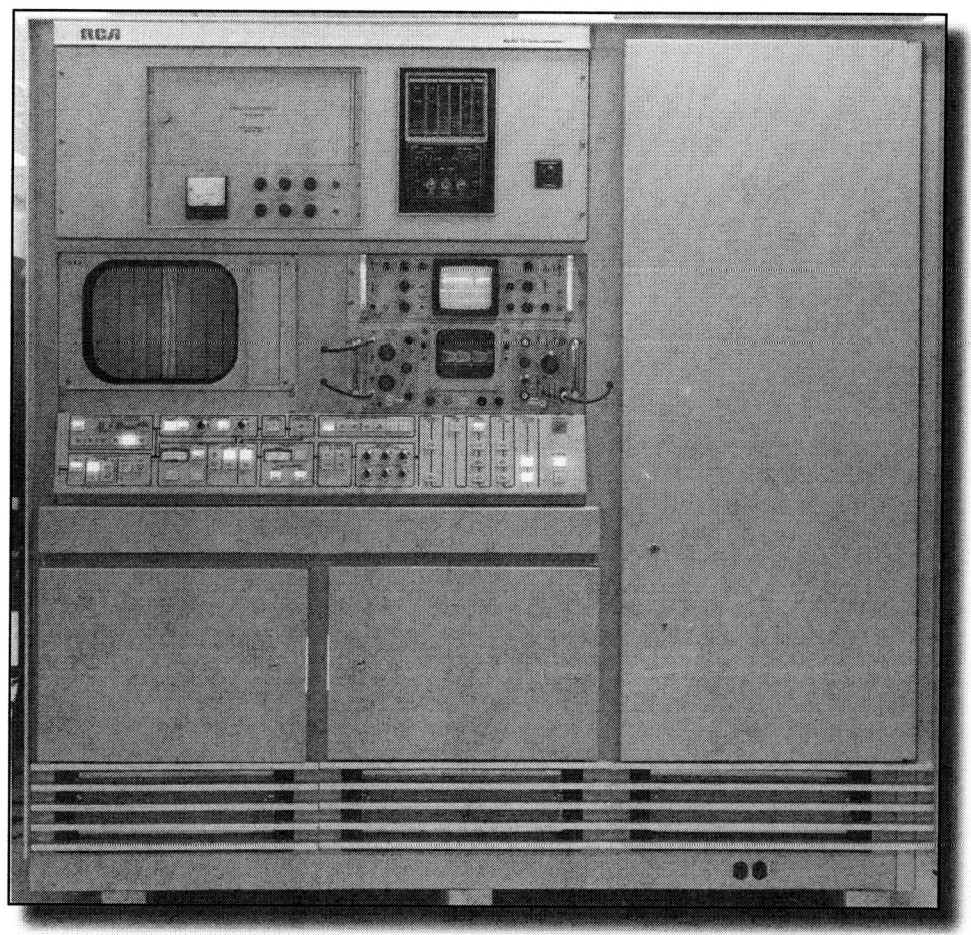

The RCA scan converter used to convert the slow scan black-and-white television images from Apollo 7,8,9 and 11. Photo courtesy of Dick Nafzger.

resolution mode, while always available, was never actually used. A variety of controls relating to video synchronization and video level settings were accessible to adjust the signal, in most cases the controls were separately configured for either the standard or high resolution modes. The final output path allowed image enhancement and adjustment of the video disk recorder. The resultant signal could be made as visually appealing as possible, although despite all the controls, the fact was that the slow scan image was optically converted - and that resulted in a loss of resolution of the picture.

A contingency plan submitted by Westinghouse contained a proposal and study to modify and adapt the slow scan black-and-white lunar TV camera to provide a color image. It consisted of a mechanical device placed over the TV camera lens it which was basically an external version of the color wheel in the color Westinghouse cameras. A reference black segment was incorporated on the wheel to allow calibration as the sequence started. Three quadrants of the primary colors red, green and blue were automatically moved into position when the starter sequence was started by the astronaut. The filters were moved by way of a spring loaded rotating wheel mechanism. Each filter would rotate into position resulting in a sequential color signal being sent via the slow scan TV camera to Earth. The movement and dwell time of each filter was 4.8 seconds with the end color, blue being the one the mechanism rested on until the rotation sequence was started again.

This color sequence would be reconstituted on the ground via a Polaroid camera with a color filter placed over its lens. A decoder system was put in place to match the ground color wheel with the appropriate incoming color. The three color frames would then be exposed on top of one another in the Polaroid camera with the resulting photograph being a full colour image! This mechanical system was devised as a back up to the color film cameras in case of malfunction, or in a worst case scenario, if the astronauts became stranded on the moon, they could at least return immediate color images of the surface features. Despite having a weight of only 1.5 pounds, the idea was never adopted by NASA.

The Westinghouse lunar television camera was ready for delivery to MSC for the flight of Apollo 9. On that flight it was given a successful, but brief operation in zero gravity. The original plan to use it in the vacuum of outer space was scrubbed due to the space sickness problems encountered by Rusty Schweickart. The stage was set for implementation onto the flight of Apollo 11; given that the last minute push to have color TV on Apollo 10 had been triumphant.

A Pre-Installation Acceptance Plan was drafted March 12, 1968 which outlined regular operation and maintenance to be performed on the camera to ensure its operational parameters had not degraded. The general recommendation was that every 90 days the camera be operated for at least one hour, and its performance assessed. Should the 90-day requirement be neglected a series of procedures involving the cleaning and checking of the camera were outlined in the guidelines. As the confidential SEC tube was housed inside the camera casing, security issues were not breached in the performing of the duties required by the Pre-Installation Acceptance Plan.

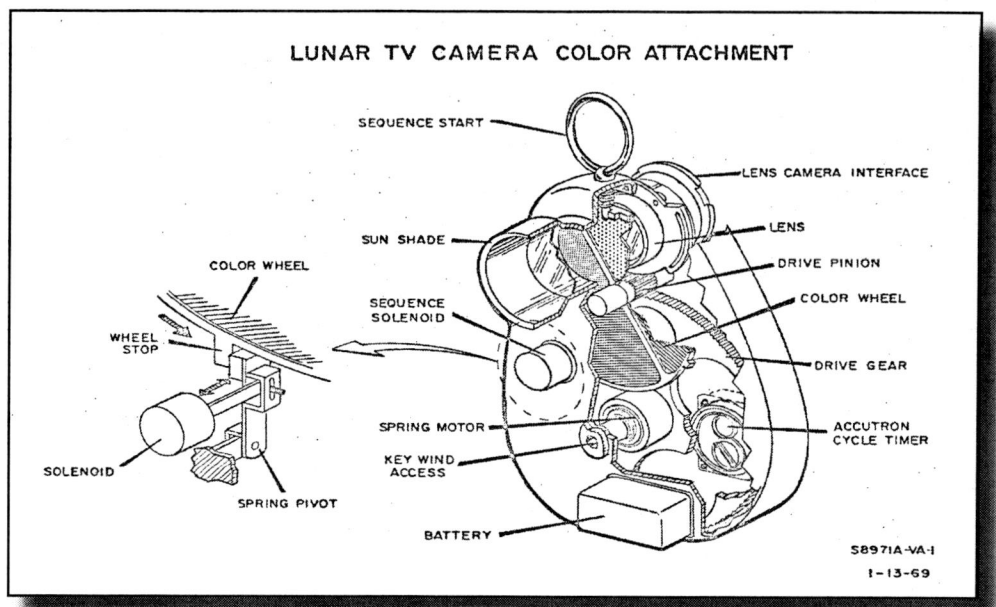
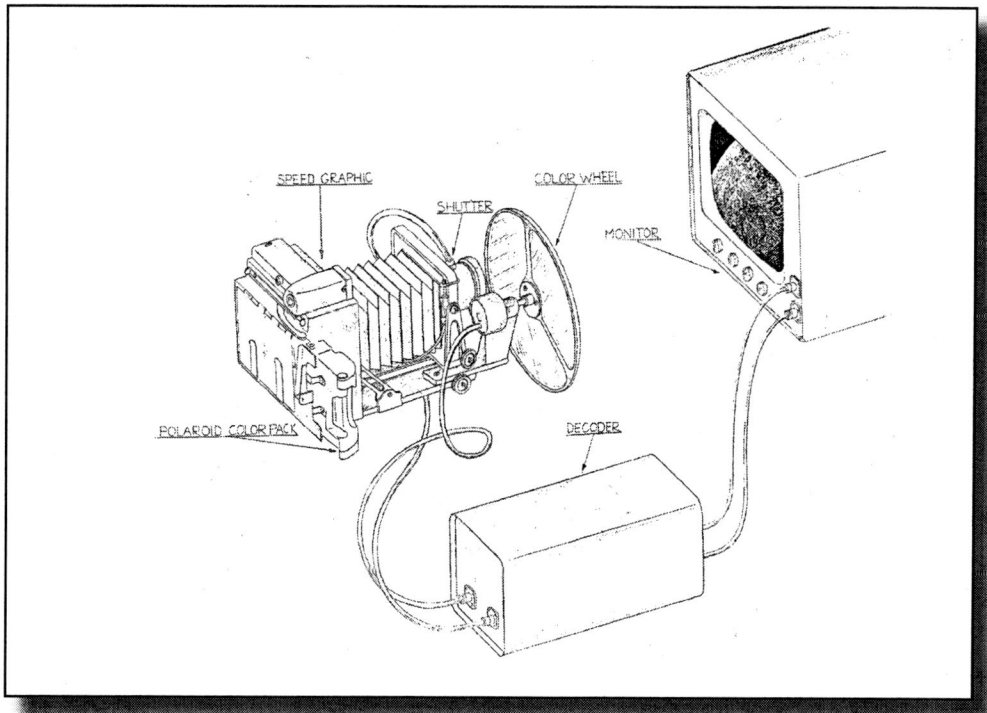

The design and tests of the Westinghouse mechanized color modification to the lunar surface camera. Photos courtesy of Stan Lebar.

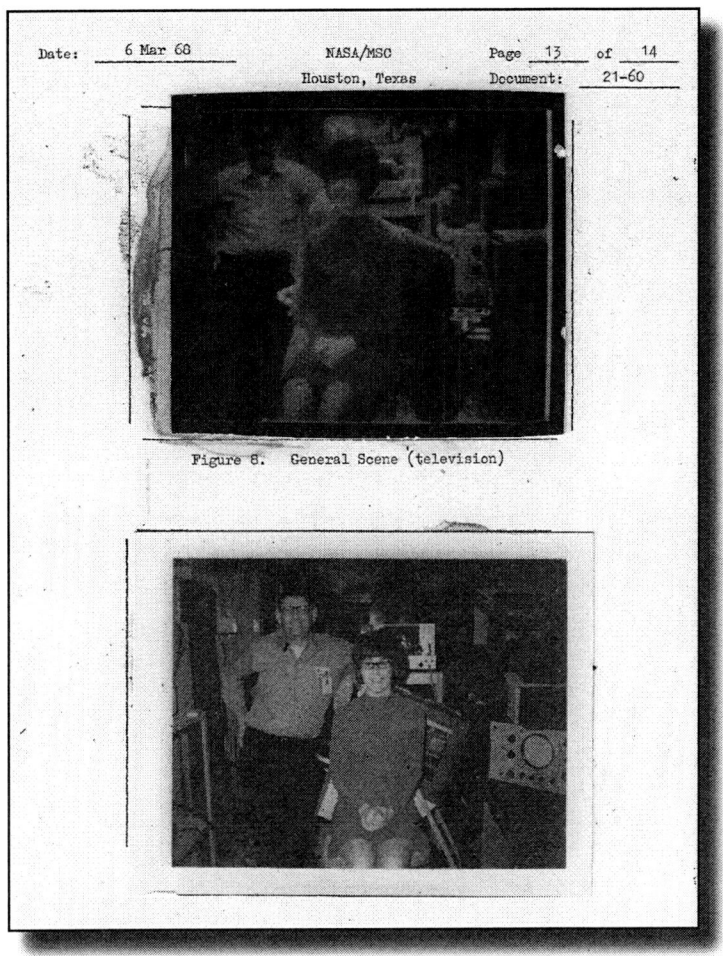

Figure 8. General Scene (television)

Discussions as to the practical uses of the TV camera on the lunar surface continued. In an ongoing struggle with the Public Affairs Office, regarding the huge publicity factor of having television from the moon, NASA management was committed to having all relevant departments once again argue their case for the inclusion of television. When considering the engineering and scientific advantages of live TV, the argument was usually referred to as having "some merit" and thus was easily set aside as unimportant to mission success. An April 15, 1968 memo by D.D. Lloyd recommended the investigation of using the Lunar Module steerable antenna to provide television prior to the planned set up of the erectable S-band antenna to record critical phases such as contingency sample collecting and most importantly egress from the Lunar Module onto the lunar surface. In many ways this requirement was as interesting to the public as it was desirable to mission planners, yet management seemed only concerned with mission critical applications of the camera during the lunar EVA.

William Hess issued a memo on July 28, 1968, wherein he agreed with the conclusion that, "…TV is not primarily required for scientific purposes." George Low responded to this on August 6 by outlining the processes in which television could assist in real-time mission planning, however he also pointed out that for the time being, "TV be considered primarily as a supplemental data source." based solely on the limited experience which had been obtained so far. However, he did recognize that as more supportive data and demonstrations of television's reliability and quality were obtained, the avenue would remain open for TV to obtain a more important role in the mission plans.

Test after test was performed to ensure the camera was fully qualified for the flight. Unlike any other component of Apollo hardware the TV camera was expected to function during any phase of the mission if so required. Westinghouse was committed to ensuring that it did. The engineers were adamant to have their lunar surface camera provide the best picture possible under all foreseeable circumstances from the moment the astronauts opened the Lunar Module hatch. The undercurrent mentality was that television was exactly what was needed to bolster public support of the space program--and what better way to do that than to record the first step onto the lunar surface?

In early 1969, a number of EVA simulations were performed in which the lighting conditions on the moon were approximated. The results of such tests concluded that there would be adequate lighting from scatter and reflection to illuminate an astronaut standing in the Lunar Module Shadow. In conjunction to this, the TV camera would be capable of transmitting a dimly lit object even if brightly lit items featured in the same scene. Additionally, two sets of stripes were painted on top of the TV camera to assist the astronauts in estimating the field of view captured by the camera when using the two types of lenses.

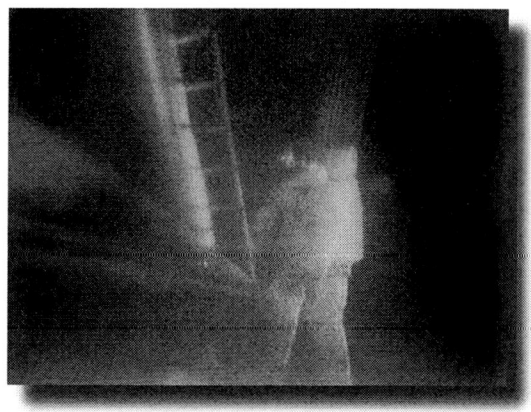

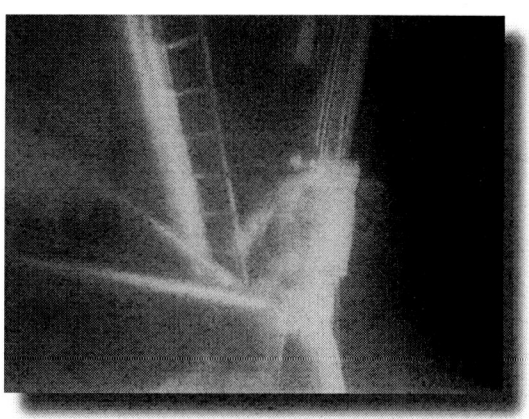

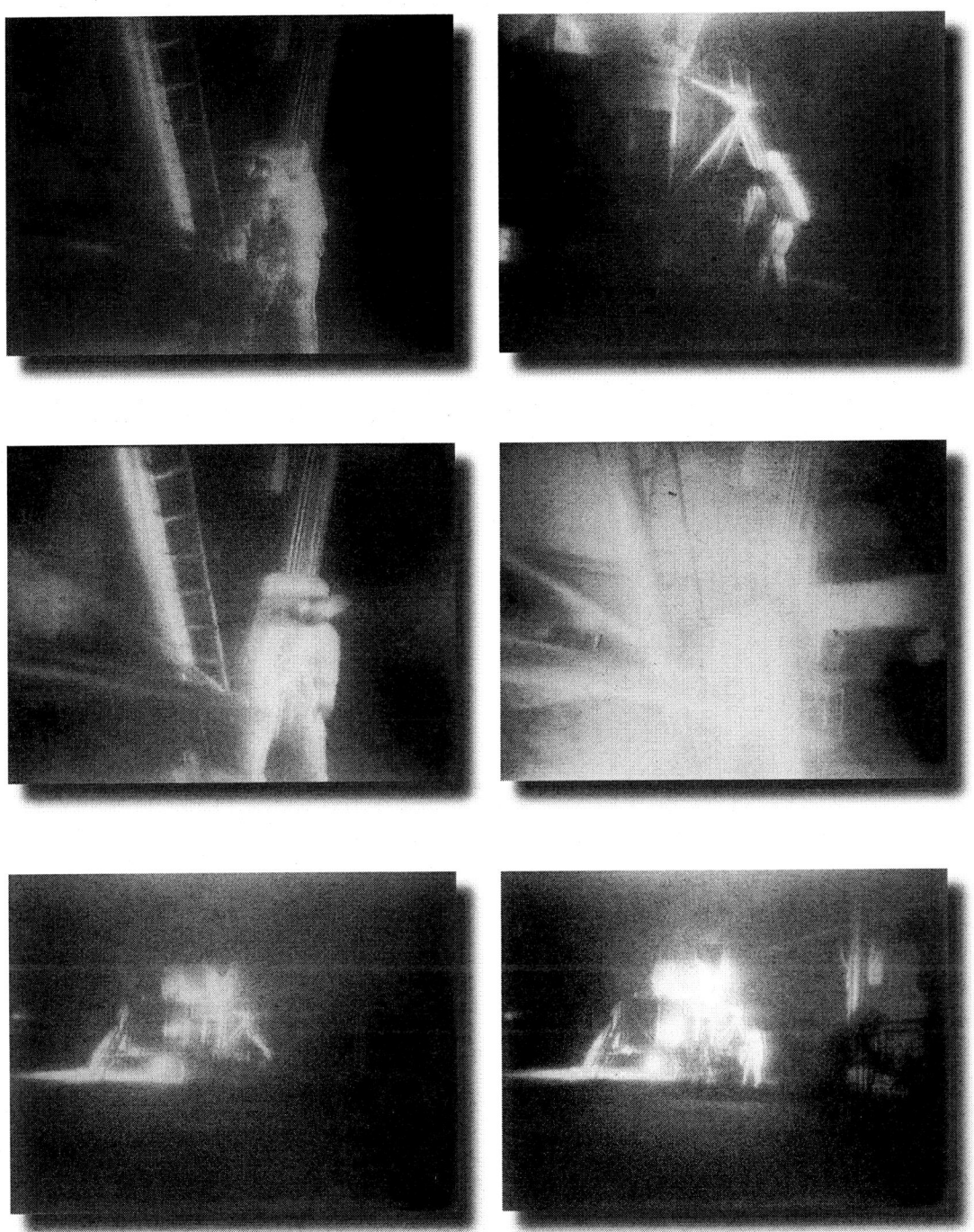

Rare photographs of a test of the lunar surface TV camera with simulated Apollo 11 video on May 7, 1969 (NASA photographs S69-33118 through S69-33126.

Rare photographs of a test of the lunar surface TV camera with simulated Apollo 11 video on May 7, 1969 (NASA photographs S69-33127 through S69-33132.

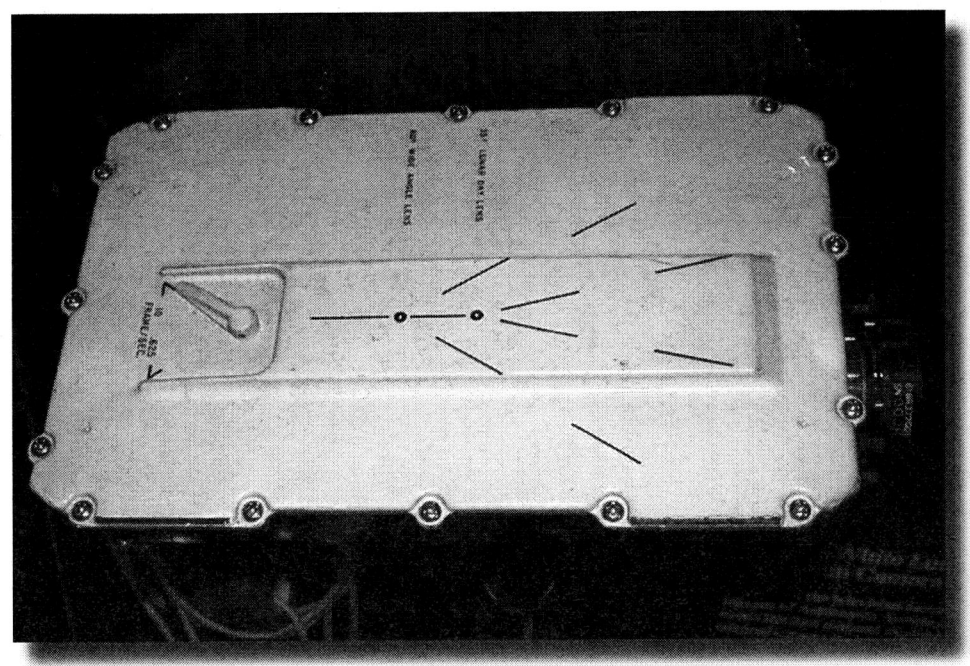

The markings found on the top of the lunar surface camera which guided astronauts in estimating the correct field of view when pointing the camera during the EVA. (Photo courtesy Ulli Lotzmann)

On June 27, 1969 less than a month before the first lunar landing attempt, NASA issued a report detailing the lunar surface activities planned. TV coverage was described as follows, "The primary purpose of TV is to provide a supplemental real time data source to assure or enhance the scientific and operational data return. It may be an aid in determining the exact LM location on the lunar surface, in evaluating the EMU, and man's capabilities in the lunar environment and in documenting the sample collections. The TV will be useful in providing continuous observation for time correlation of crew activity with telemetered data, voice comments, and photographic coverage."

Shortly after the successful Apollo 10 mission the question of television being used on the lunar landing mission became the focus of discussions at NASA. The scientists who had planned experiments to be conducted on the moon seemed unconcerned whether TV was beamed back to earth or not. They had no specific objection to it, but they couldn't rationalize inclusion of it either, prompting George Low to insist that the matter be resolved. According to Chris Kraft who oversaw flight operations, Ed Fendell made a presentation at an organized meeting regarding the Apollo 11 mission, in which he flatly recommended TV not be taken to the moon! In defiance of what they felt to be a short-sighted recommendation, Chris Kraft, Max Faget and Julian Sheer successfully argued for the inclusion of the Westinghouse camera on the historic mission. The American taxpayer had the right to observe the mission their tax dollars had financed, and those who supported the inclusion of television wasted no time in stating this important fact to those in opposition.

(NASA Photo S69-31575)

The camera was to be stored in the Modularized Equipment Stowage Assembly in a manner which pointed straight at the ladder attached to the landing leg of the spacecraft. Upon opening of the MESA, the camera would be ready for activation to shoot the Commander's descent onto the lunar soil.

Following the descent of the Lunar Module pilot onto the surface, the camera would be removed, placed on a tripod, and placed in a position which monitored the Lunar module and the astronauts' activities in the immediate vicinity. Depending on the ability of a 210 foot antenna to receive the steerable antenna signal and its associated TV picture, the deployment of the erectable antenna may not have been required.

As this procedure required 19 minutes of surface time, it was hoped to be unnecessary.

The operation of the camera was expected to be successful the moment it was activated. There was no other alternative – no second chance. In effect, the turning on of the camera would be the ultimate test of the television system as noted in the Westinghouse Engineer article from 1968:

"The final test of the lunar TV camera will be viewed by millions as they watch, on their home television sets, the exploration of the moon as seen by the Apollo astronauts."

Had the erectable S-Band antenna been used, this training photo shows how it would have been deployed by the astronauts. (NASA Photo S69-31163)

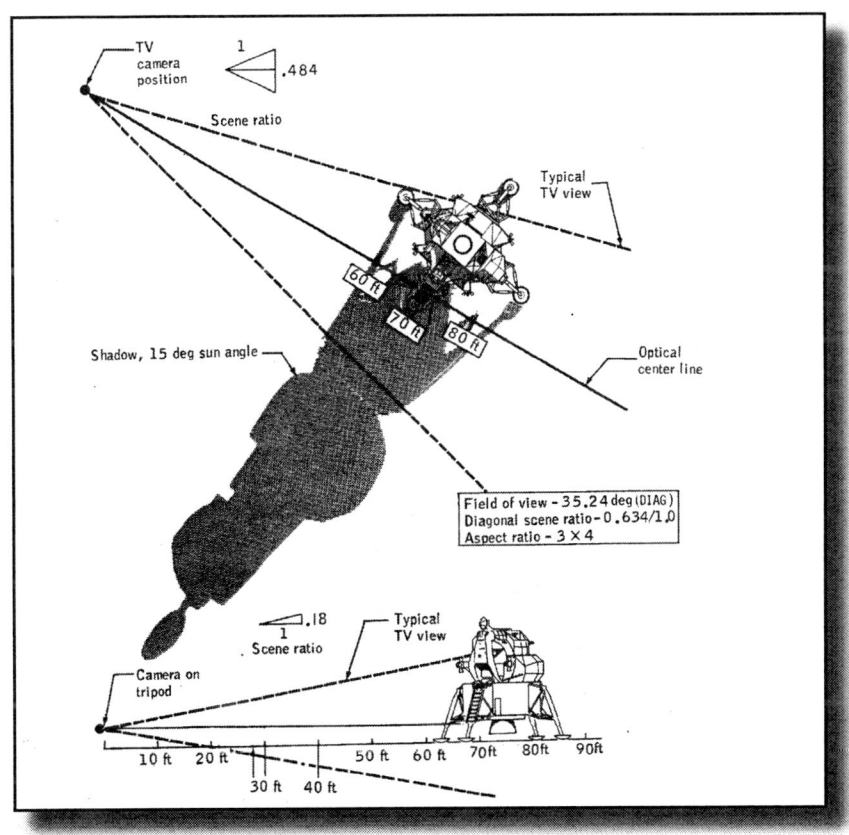

Daytime lens TV field of view from tripod.

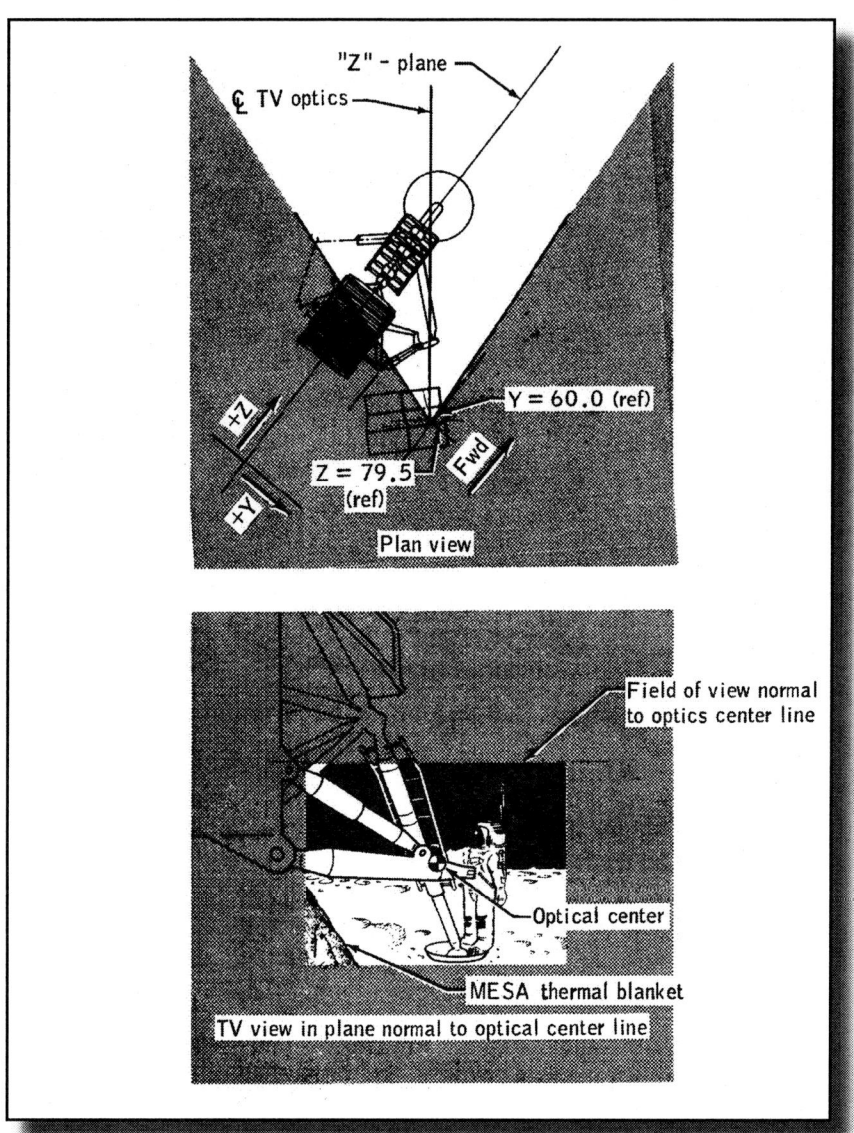

TV field of view from the MESA.

The flight of Apollo 11 would be one of the most anticipated news events of the century. Numerous delegations of journalists were in the United States to coverage the start of the mission from Florida. Countless others were based in their homelands watching satellite transmissions of the Saturn V launch which would carry three astronauts to the moon. Be it radio, print or television media, the activity around the Apollo 11 mission was huge. It would be TV cameras which brought the mission into the homes of the average citizen watching from the comfort of their armchairs. This time, it didn't matter if their favourite programs were interrupted. It was history in the making and people wanted to watch it live.

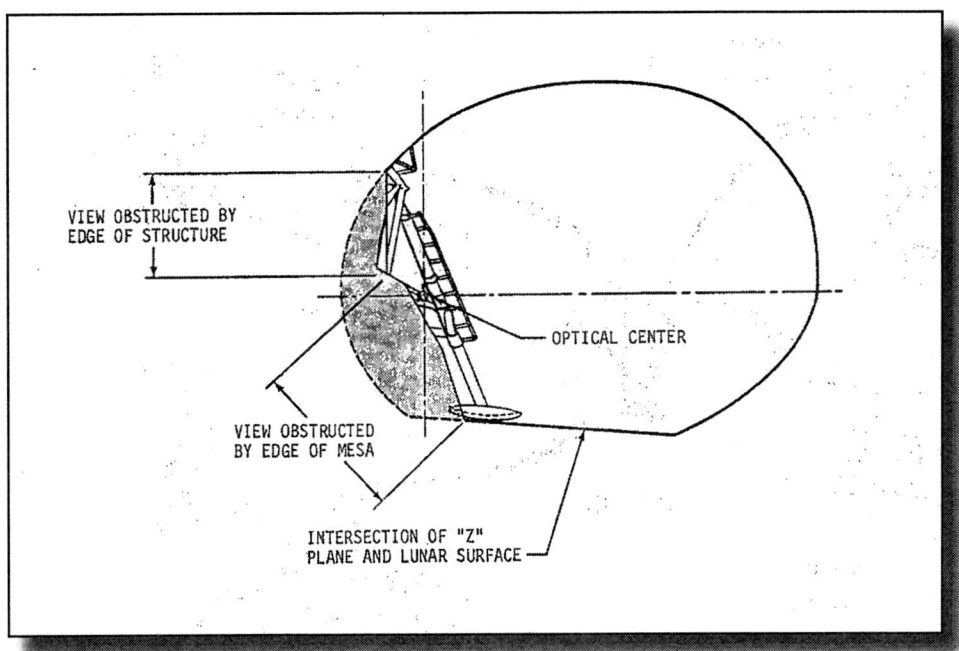

View through optical center of TV lens in direction of "Z" plane.

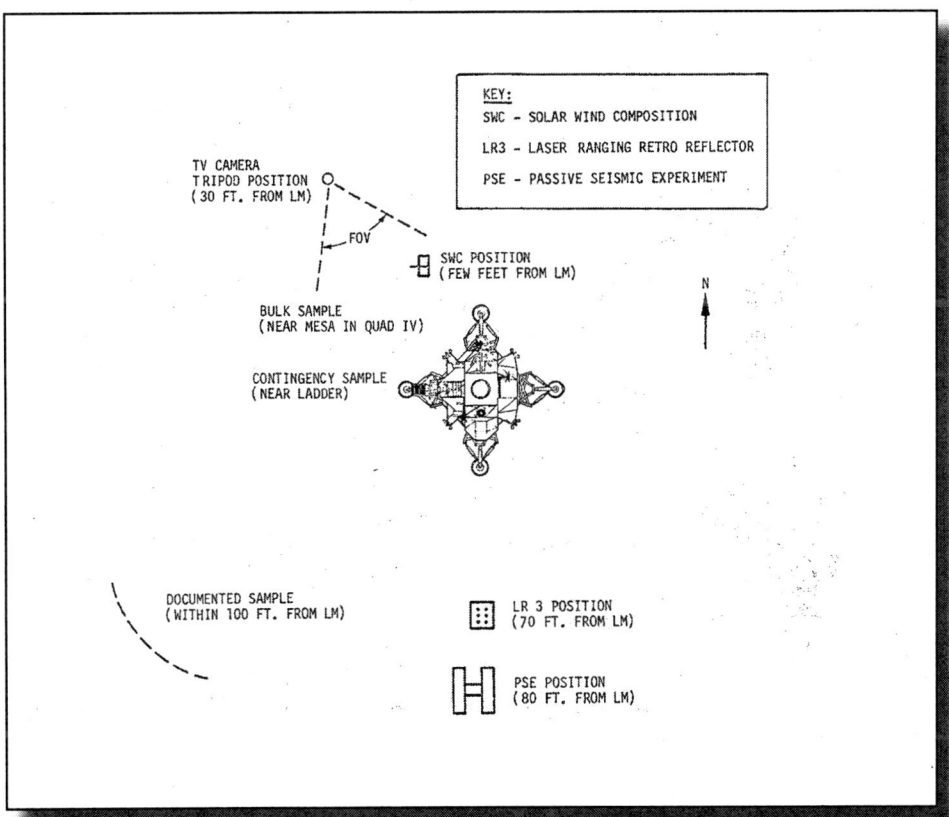

The planned setup location for the TV camera once it was removed from the MESA.

Apollo 11 launched on July 16, 1969 with Neil Armstrong as Commander, Mike Collins as Command Module Pilot, and Buzz Aldrin as the Lunar Module Pilot. The excitement felt by almost everyone extended all around the world. From Australia to inside the Iron Curtain, the feed of the launch was shown to millions of eager viewers. As the Saturn V rose off its launch pad, there was a genuine sense of wonder. It wasn't a dress rehearsal or an Earth-orbital mission - these men were not only going to the moon, but two of them were actually going to walk on it.

An initial indication that the color TV would be activated on the first revolution of the spacecraft around the Earth was not followed through by Apollo 11's crew. The first television transmission was made a little over 10 hours into the flight with a view looking back at the blue ball of the earth. It was unscheduled, and unfortunately Houston was unable to see the magnificent sight as Goldstone received and recorded it. Interior cabin views showed Neil Armstrong hanging upside down fast asleep. Russian news reports had named him the "Czar of the Ship"! Mike Collins was actively looking through the optical sights, while Buzz Aldrin gave a silent tour of the ship.

A second unscheduled TV transmission was made 20 hours later showing the DSKY computer display, although rather distorted through a bad signal. The camera was turned on to test the equipment and the crew informed Houston, who then contacted Goldstone at the request of Mike Collins to verify that they could read the numbers from the computer display. The recorded signal was indeed startlingly clear although horizontal interference made the images difficult to watch, appearing similar to bad reception of a distant channel on a home TV set. Collins demonstrated how the astronauts maintain some form of physical activity by running on the spot, bringing the flight surgeon to life at his desk in Mission Control thanks to the increase in heart-rate!

At 33:59 hours after launch, the crew began the telecast which was supposed to be the first one scheduled. Beginning with the earth shot from the left-hand window of the spacecraft, the astronauts gave a travelogue of what they could see upon looking down at the planet. When the camera trained on the men in the spacecraft, Mike Collins was once again clearly visible. He then proceeded to give a small tour of the Command Module, and jokingly referred to the Apollo 7 crew's use of cue cards by saying, "we have no intention of competing with the professionals." Buzz did some short zero-G exercises, and for the benefit of TV viewers Collins demonstrated how chicken stew is prepared on a long journey to the moon. Charlie Duke conveyed the thoughts of everyone watching by saying, "Apollo 11, Houston, we're really impressed with the clarity and the detail we have in this picture...it's really an excellent picture that I'm looking at on our monitor..."

An interesting segment displaying the computer system was given by Mike Collins, in which, after warning the flight controllers back in Houston, he entered a command which lit up the entire display. In color it was spectacular. The transmission then concluded with the view of Earth once again through the spacecraft window.

Buzz Aldrin setting up the 16mm camera inside the LM.

The next transmission was the familiar probe and drogue inspection and removal. Without much commentary the crew examined the connection between the Command and Lunar Module and while the color pictures displayed a startling amount of detail, the television feed was primarily for the engineers at Houston. The transmission certainly lacked the "guided tour" feel of the previous telecast. That was until they entered the dark Lunar Module. For millions of people on the Earth, this was the first view of the craft that would actually land on the moon.

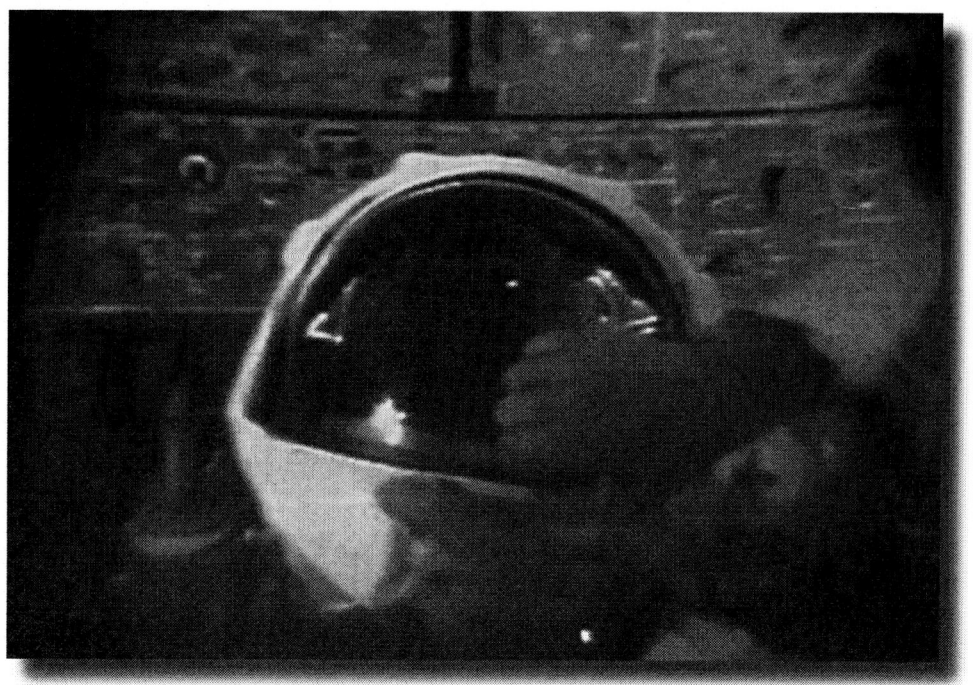

Buzz demonstrating the spacesuit helmets to the chagrin of Wally Schirra watching in the CBS TV studios back on earth.

For the astronauts it was also their first inspection of it during the flight so while the camera was running, they quickly examined the stored items which would be used on the lunar surface. When Buzz Aldrin picked up the helmet to demonstrate how the visors worked, thus possibly contaminating it, Wally Schirra, who was co-anchoring the CBS coverage of the mission commented, "There's an interesting facet to this. There's no quality control inspector to chew him out for touching this thing with his bare hands. He should have gloves on." While live television allowed people to share in the glory of the lunar mission, it also afforded them the ability to pick up on mistakes made by the astronauts in their spacecraft several million miles out in space as they happened!

As the two spacecraft entered Lunar Orbit, and flew over the planned landing site, they turned on the TV camera to offer close views of the moon. They described, as had their counterparts on Apollo 10, the surface features and craters as they passed below. The anticipation was certainly mounting as the moment which had been in development for nearly a decade had almost arrived. All that remained now was to undock the Lunar Module and land it on the surface.

Unfortunately, the landing of the Lunar Module was not televised, although the idea had been suggested to NASA. A memo issued on April 27, 1964 raised the issue of television during the spacecraft descent to the moon in which Dr Lee, Chief of the Operations Planning Division stated, "It may be assumed that the first attempt to land on the moon will have generated a high degree of interest around the world. The public will have witnessed the launch and followed the progress reports avidly as the spacecraft travelled toward the moon. Undoubtedly, there will have been much speculation during the long flight out about the probability of a successful landing. A large portion of the civilized world will be at their TV sets wondering whether the attempt will succeed or fail. The question before the house is whether the public will receive their report of this climactic moment visually or by voice alone." Lee further suggests, "It is, however, important to recognize that a critical component of the public interest is the suspense generated by uncertainty over the outcome of the landing attempt. This suspense will not be present a few hours later when the astronauts erect an antenna on the surface, nor will it be present when films are available after the return to earth." Sadly, for the viewing public, no sooner had the recommendation been made, it was stopped dead in its tracks. No further investigation was made, due partly to the lack of necessary equipment to make the suggestion feasible and partly due to the view that television had no important data value.

As the mission profile dictated, the first time television would be received from the moon was shortly before the astronauts were about to stand on its surface. The TV camera had changed its location several times during the mission development. Originally, it was to be transferred from the Command Module into the Lunar Module and taken to the surface inside the spacecraft. The ongoing weight revision program of the LM ascent stage resulted in a further relocation of the camera to the descent stage location of the Modular Equipment Storage Assembly (MESA) which was completely outside of the spacecraft. The camera was in a position which focused directly on the LM ladder once the MESA was opened via a lanyard system, pulled by Neil Armstrong as he was coming down to the surface.

Despite having a seat on the flight, television was not given much thought or priority by NASA in the months prior to the landing attempt of Apollo 11. Deke Slayton made his feelings known on September 16, 1968 regarding the use of the larger tracking station antenna. He stated, "The EVA is planned to start at essentially a fixed time from Lunar Module touchdown and is not based on the availability of the 210 foot antenna for TV reception. The use of the 85 foot antenna would be made only after a preliminary test of the image. Indeed, TV would have to wait until the proper receivers were in line of sight for the TV signal. Luckily such plans did not eventuate; there were several antennas ready to take the television pictures from the moon regardless of what time the EVA occurred after landing.

This turned out to be a good thing for all those interested in watching the first moonwalk. Shortly after landing, the crew requested that they forego their scheduled rest period and start the EVA 4 hours earlier than planned. Parkes would not yet be in sight of the Lunar module when the moonwalk would start, and so Goldstone became the prime receiving station. All ears and eyes were trained upon the communications coming from the moon.

As Armstrong was ready to exit the Lunar Module, Aldrin gave a brief pep-talk, "Okay. About ready to go down and get some Moon rock?" The egress was time consuming given that the astronaut would need to crouch down and slowly manoeuvre through the hatch, guided by his companion. As Armstrong passed the D-ring lanyard cable which would deploy the MESA containing the TV camera, he grabbed hold of it.

Aldrin: "Did you get the MESA out?"

Armstrong: "I'm going to pull it now. Houston, the MESA came down all right."

McCandless: "This is Houston. Roger. We copy. Standing by for your TV."

Before doing anything else, Armstrong called to Houston for a radio check, the final verification that all communication would be heard as he stepped onto the moon.

Armstrong: "Houston, this is Neil. Radio check."

McCandless: "Neil, this is Houston. Loud and clear. Break. Break. Buzz, this is Houston. Radio check, and verify TV circuit breaker in."

Aldrin: "Roger, TV circuit breaker's in. And read you loud and clear."

McCandless: "Roger."

Back in Mission Control, the large eidphor video-projector screen displayed a test pattern of grey-scale bars. The world sat on the edge of its seat as it anticipated the first live television of men on the moon. The flurry of activity behind the scenes was concealed by the relative calm of the astronauts and Mission Control. 16 seconds prior to the television feed, when asked to confirm a signal from the TV camera, Goldstone reported to Houston, "Roger, we do have FM downlink, we are not receiving TV at this time." Had there been a problem? Was the camera operational? There was too little time to think.

Honeysuckle was the first to report something appearing on its receiver, "Copy possible TV OPS", indicating that there was a signal coming in, more than likely a TV picture, but no definite confirmation as yet. Then, 12 seconds later Goldstone called out jubilantly, "TV online. Goldstone TV online!" Immediately following, Honeysuckle also confirmed with a picture with the same amount of euphoria, "Honeysuckle video online!" the display in Houston sputtered to life with an image, confirming that a television signal was arriving from the moon.

McCandless: "And we're getting a picture on the TV!"

Aldrin: "You got a good picture, huh?"

McCandless: "There's a great deal of contrast in it; and currently it's upside-down on our monitor, but we can make out a fair amount of detail."

Despite getting a television feed from the moon, what Bruce McCandless reported was not good. The scan converters had been calibrated to deliver a reasonably good image, not as sharp as the slow scan image directly from the camera, but definitely not as degraded as the image seen at Houston. Something was not right, and it needed to be rectified quickly. The step onto the lunar soil was only moments away. Houston asked Goldstone, "Can you confirm that your reverse switch is in the proper position for the camera being upside down?"

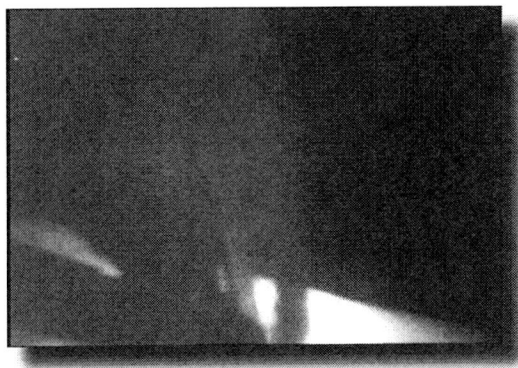

Armstrong on the ladder as seen on the TV feed from Honeysuckle Creek and broadcast only in Australia.

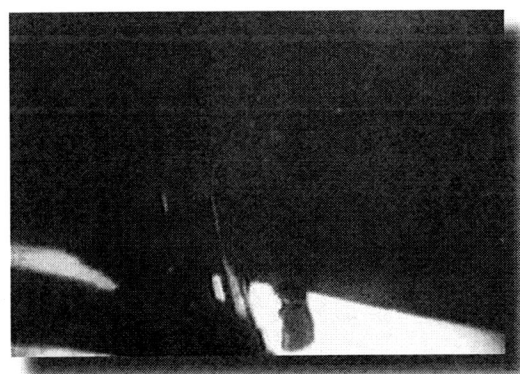

The image quality of Armstrong's descent as sent to NASA from Goldstone.

In the rush of activity prior to the arrival of the television signal, someone at Goldstone had inadvertently selected the toggle switch for the camera when in its upright position. Analysis of the feed from the tracking station also suggests that the controller operating the scan converter simply panicked when they saw the upside down picture and began to compensate by changing the brightness and contrast settings. It only made matters worse. Ironically, as the Goldstone antenna was larger than Honeysuckle's, the picture quality would have been superior had the scan converter settings been left alone.

Aldrin's comment regarding a good picture established whether Armstrong would need to deploy the erectable S-band antenna to improve the signal quality. During training exercises, the antenna setup had been performed. However, it did take several minutes to do. An April 27, 1966 memo from R.L. Seldon reported that the use of the LM steerable antenna would deliver useable voice and television. By not having to erect the antenna, the astronauts would have more time on the surface to conduct research and experiments.

McCandless: "Okay. Neil, we can see you coming down the ladder now."

The Goldstone images were processed through the scan converter and then sent directly to Houston. The feed was then routed to the television networks all around the world. In the United States, the images came from Goldstone via Houston, and had a .3 second delay built into the feed due to sending the signal via microwave, and satellite from the tracking stations to Houston and then onto the world's television stations. Unbeknownst to them, since the start of transmission from the Lunar Module, Australians were able to view the moonwalk a fraction of a second before the rest of the world as they received the feed straight from Honeysuckle. Australians also had the benefit of a properly set scan converter giving them their television pictures. There were no crushed brightness levels and no upside down picture in that part of the world!

Seeing that Goldstone was having problems with their image, Houston decided to switch to Honeysuckle just seconds before Armstrong put his left foot onto the lunar surface. Each time a switch was made, the video synchronization was temporarily lost, resulting in a break in the picture's vertical and horizontal hold. While noisier than the Goldstone feed, it had the correct brightness levels, and Armstrong was plainly visible in the frame. Now, in just a few moments, centuries of dreaming and a decade of hard work were about to be realised. While actually landing on the moon had been a triumph in itself, the act of putting a human foot onto the surface was the icing on the cake. Humanity was about to enter a new era.

Armstrong: "I'm going to step off the LM now."

Neil Armstrong still held on to the ladder of the Lunar Module with his right hand. With his right foot also well within the footpad of the Lander, he placed his left foot onto the surface of the moon.

Armstrong: "That's one small step for man; one giant leap for mankind."

The whole time the television sent its signal through the Lunar Module's antenna to the earth. Millions of people around the world watched as the historic first step was made. Armstrong had intended to say "One small step for a man; one giant leap for mankind," but in the excitement of the moment he forgot to say it. Over the years the omission has been subject to much discussion, yet there is little to indicate the "a" was clipped by the communication circuitry. The vowel was simply forgotten.

The mission commander continued his examination of the immediate vicinity around the Lunar Module. After being handed down the Hassleblad lunar surface photographic camera, he picked up a contingency sample of the lunar soil. The moonwalk of Apollo 11 differed markedly from the others in that it was cautious by nature. Despite being on the moon for nearly ten minutes, Armstrong was not prepared to venture more than 10 feet until he was comfortable in doing so. Approximately 20 minutes after the first step, Buzz Aldrin descended the ladder and joined his colleague on the lunar surface. The TV at this point was now coming from Parkes and the clarity of the picture was vastly superior to either Goldstone or Honeysuckle. Finer details of Aldrin's suit could be made out which was not the case on his colleague's first moments in front of the camera.

Armstrong moved to the TV camera to change its lens while Aldrin walked on the moon's surface acclimatizing himself to movement in 1/6th gravity.

Armstrong: "Okay, Houston. I'm going to change lenses on you."

McCandless: "Roger, Neil."

The black-and-white lunar surface TV camera was not designed with a zoom lens. Since the beginning of development the use of detachable lenses was planned, and at this point in the EVA Armstrong was to change the wide angle lens, necessary to view the ladder of the Lunar Module, to a normal angle lens to view the EVA from a distance away from the LM.

Armstrong: "Okay, Houston. Tell me if you're getting a new picture."

McCandless: "Neil, this is Houston. That's affirmative. We're getting a new picture. You can tell it's a longer focal length lens. And for your information, all LM systems are Go. Over."

Aldrin: "We appreciate that. Thank you."

The astronauts now proceeded to unveil a plaque which was attached to the ladder on the Lunar Module. All subsequent missions would have a plaque attached to the spacecraft, but the one for Apollo 11 was historically significant.

Armstrong: "For those who haven't read the plaque, we'll read the plaque that's on the front landing gear of this LM. First, there's two hemispheres, one showing each of the two hemispheres of the Earth. Underneath it says "Here Men from the planet Earth first set foot upon the Moon, July 1969 A.D. We came in peace for all mankind." It has the crew members' signatures and the signature of the President of the United States."

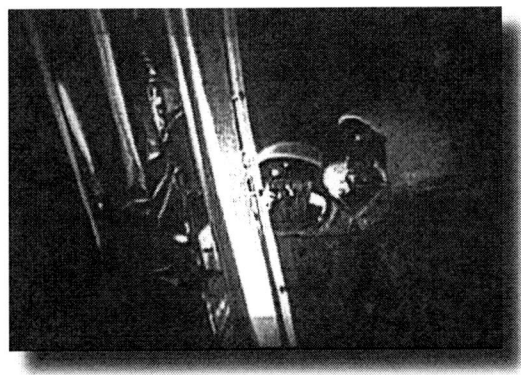

Prior to the removal of the TV camera from the MESA, Buzz Aldrin held his gloved hands up to it to show the effects of the lunar dust on his spacesuit. Aldrin then proceeded to supply the television cable to Neil as he moved the camera to its new resting position.

The lunar surface swung by in a blur. On the television camera was a color patch which had a series of dots on it which would change from white to black as the temperature increased. Armstrong had reported that the camera's temperature was cold after he inspected the patch. Aldrin speculated that the TV camera was between 40 and 50 feet from the Lunar Module, and suggested Armstrong turn the camera around to show the view of the landing site to Houston for evaluation. The loose TV cable posed somewhat of a nuisance as it easily was caught between the astronauts' boots, and Aldrin recommended Armstrong turn to his right, which also happened to be in the direction of the sun. The black-and-white TV camera did not have an adjustable iris and it also had an ultra-sensitive pickup tube. Pointing it at the sun would be fatal for the camera.

>Armstrong: "I don't want to go into the Sun if I can avoid it."

>Aldrin: "That's right, Yeah."

Following a few directions from McCandless, the camera was placed in position pointing directly at the Lunar module. Mission Control wrongly assumed Armstrong was making a panorama of the landing site as he was trying to locate a decent spot to leave the camera. The mission geologists in Houston would take Polaroids off the TV screen to assist them in making assessments of the area and obviously a blurred picture was of no use to them. Once suitable stable ground had been found, Armstrong panned the camera at set intervals for the geologists giving a brief description of what he was pointing at. Once that had been completed the camera was pointed again at the Lunar Module, where it would stay pointed for the remainder of the mission.

Another symbolically important part of the mission now took place as the two men raised the American flag on the surface of the moon, and in full view of the television camera. The nylon flag was attached to a horizontal rod which held it into place in the airless environment. As the rod was not properly extended, the flag remained somewhat crumpled, giving the impression it was fluttering in the wind. The flagpole it was mounted to did not easily go into the ground. A great deal of effort was required to have it in the surface far enough so as not to topple over. As the two astronauts saluted the flag a telephone call was arranged between Houston and the White House.

>McCandless: "We'd like to get both of you in the field-of-view of the camera for a minute. Neil and Buzz, the President of the United States is in his office now and would like to say a few words to you. Over."

>Armstrong: "That would be an honor."

>McCandless: "All right. Go ahead, Mr. President. This is Houston. Out."

>Nixon: "Hello, Neil and Buzz. I'm talking to you by telephone from the Oval Room at the White House, and this certainly has to be the most historic telephone call ever made. I just can't tell you how proud we all are of what you have done. For every American, this has to be the proudest

day of our lives. And for people all over the world, I am sure they, too, join with Americans in recognizing what an immense feat this is. Because of what you have done, the heavens have become a part of man's world. And as you talk to us from the Sea of Tranquility, it inspires us to redouble our efforts to bring peace and tranquility to Earth. For one priceless moment in the whole history of man, all the people on this Earth are truly one; one in their pride in what you have done, and one in our prayers that you will return safely to Earth."

Armstrong: "Thank you, Mr. President. It's a great honor and privilege for us to be here representing not only the United States but men of peace of all nations, and with interests and the curiosity and with the vision for the future. It's an honor for us to be able to participate here today."

Nixon: "And thank you very much and I look forward...All of us look forward to seeing you on the Hornet on Thursday."

Aldrin: "I look forward to that very much, sir."

to work as they were already behind schedule. They unloaded the Early Apollo Science Experiments Package (EASEP) and the Lunar Ranging Retro-Reflectors. The reflectors would allow laser light targeted from earth to accurately detect the distance of the moon. To this day the reflectors are still routinely used by a dedicated facility at the MacDonald Observatory in Texas. On this mission, just as all the others, time was a limited commodity. The two men worked feverishly to complete all the tasks required of them during their allocated moonwalks. No sooner had they completed all the responsibilities, they started to end the EVA and return inside the Lunar Module.

Buzz Aldrin was the first one back inside. The rocks and lunar samples had been placed inside the cabin and he was there to receive more items from Armstrong who would still remain on the surface for several more minutes. Once everything had been properly stowed, the mission commander leaped up the ladder and he returned to the safety and comfort of the spacecraft.

The two astronauts quickly got back

Once the two crewmen had attached their life-support to the Lunar Module system, they dumped their PLSS gear onto the surface. The TV camera recorded the backpacks as they fell down from the hatch onto the moon. After the equipment dump, there was no further activity to be seen on the television image. Before they got some well deserved rest the TV camera was turned off.

Aldrin: "And, Houston, Tranquility. Have you had enough TV for today?"

McCandless: "Tranquility, this is Houston. Yes, indeed. It's been a mighty fine presentation there."

There was still ample battery power to continue running the television camera, but as there would be no change in scenery until launch, there was no further need to have the camera on. Additionally, directly upon launch, a guillotine system would sever all connections between the ascent and descent stages of the Lunar module. Had the television been left on in that configuration, the signal would be lost the moment the connections would be severed.

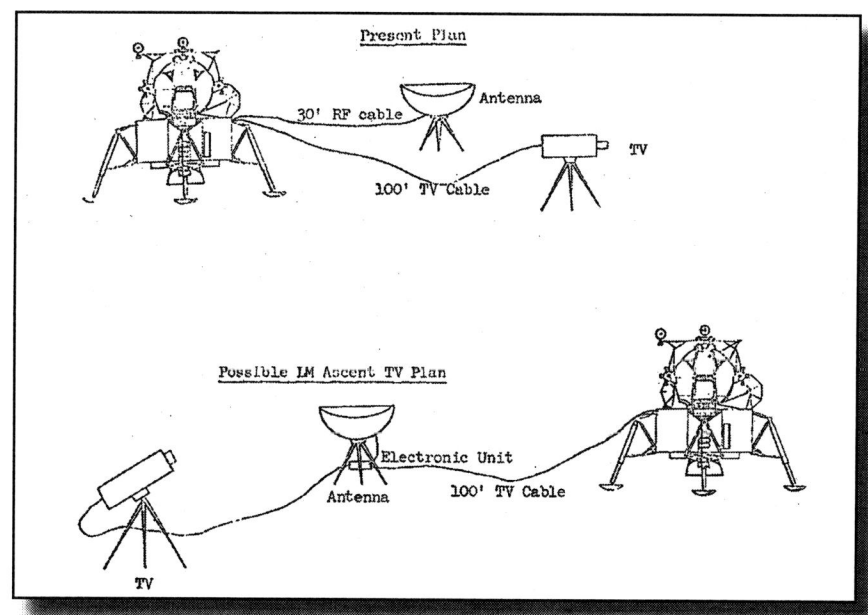

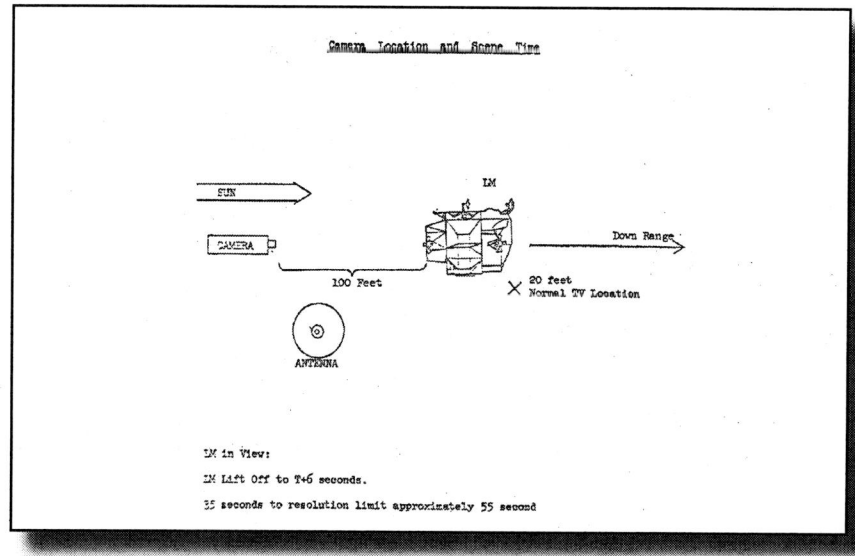

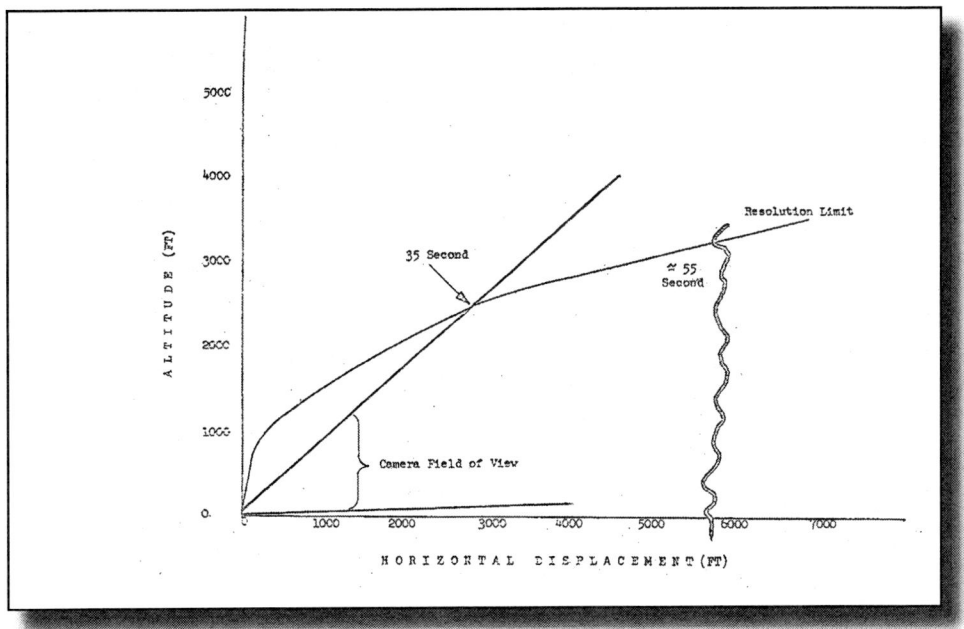

Abandoned proposed setup for the televising of the Apollo 11 LM launch.

A depiction of the launch as TV audiences may have seen it had the camera been set up to supply the earth with a transmission of the launch on Apollo 11. Panorama courtesy of Don Davis

Paul Haney had proposed the idea of televising the lunar launch as far back as 1967, comparing the amount of both engineering and public relations coverage a Saturn rocket launch had. He stated that NASA ought to make some effort in providing a view of the unique lift-off from the moon. The excitement of observing the launch in real-time via television was not lost on NASA management. Maxime Faget suggested, in a very late proposal dated March 19, 1969, a very novel, yet feasible method to televise the launch. He wrote, "It appears the most promising approach is for a new astronaut-deployed package containing a battery pack, 10-Watt S-band transmitter, and control circuitry. Activation of the system would be controlled from the ascent stage via hard-line." The memo further states, "The approach we propose would require the deployment of the LM TV camera and erectable antenna to a new position which is a deviation of present Apollo procedures." The additional weight was estimated at 15 pounds taking up approximately ½ cubic foot at a total cost of $40,000. After careful review, which had been quite enthusiastic about the approach, the idea was halted in a response by George Low in which he stated after viewing a demonstration of lunar surface procedures, "I don't think that there is any possibility of carrying the lunar ascent TV package on the Apollo 11 flight." Citing the time it would take to set up the equipment and the possible missed opportunity in obtaining precious lunar samples, he closed the matter

The world had to be content listening to the launch via voice transmissions only. In perhaps the most nail-biting moment of the entire mission, the Lunar Module launched successfully, ending mankind's first visit to the moon. There had always been a small amount of doubt in the back of people's minds that the spacecraft would not properly fire its engines and allow the astronauts to rejoin with the Command Module. Mike Collins admitted that of all the problems he imagined could go wrong, the idea of him having to leave his crewmates on the moon and fly back to Earth alone, was the one that plagued him the most. However, as the two moonwalkers inside their trusty Lunar Module approached the Command Module, his fears were finally put to rest.

A television transmission from the Command Module was made during the journey home. Beginning with a view of the receding moon, Capcom Charlie Duke mistakenly assumed it was the Earth, commenting that he could make out some land masses!

Armstrong: "Are you picking up our TV signals?"

Duke: "That's affirmative. We have it up on the Eidophor now. The focus is a little bit out. We see the Earth in the center of the screen. Still have a little white dot in the bottom of the camera, apparently. And see some landmasses in the center; at least I guess that's what it is. It's very hazy at this time on our Eidophor. Over. Let me change the screen now."

Collins: "I believe that's where we just came from."

Duke: "It is, huh? Well, I'm really looking at the bad - at a bad screen here. Stand by one. Hey, you're right."

Collins: "It's not bad enough not finding the right landing spot when you haven't even got the right planet!"

Duke: "I'll never live that one down."

Collins: "We're making it get smaller and smaller here to make sure that it really is the one we're leaving."

Duke: "All right. That's enough you guys."

Not surprisingly the crew was in remarkably good spirits. They had just achieved the goal set by an ambitious young president several years earlier. Buzz Aldrin proceeded to display some of the food the crew could eat while flying a mission. He also demonstrated the principles of torque on a spinning object in zero-G. Mike Collins joined in with the demonstrations showing a spoonful of water that just refused to spill over, even when held upside down. The show closed with a view of the approaching Earth, although this time Duke refused to bite. The crew explained what they were shooting. It was indeed the good earth.

The final television show from the historic Apollo mission was a series of crew statements in which the three astronauts discussed their thoughts on the factors which had resulted on the mission being so successful. Due to transmission problems the TV was lost periodically throughout the last telecast, while the voice channel remained operational. Houston was able to advise the crew when the television signal was back to full strength. The statements were profound and humbling. The astronauts were fully aware that it was only through the combined efforts of over 400,000 people that they were able to succeed.

Armstrong: "Good evening. This is the Commander of Apollo 11. A hundred years ago, Jules Verne wrote a book about a voyage to the Moon. His spaceship, Columbia, took off from Florida and landed in the Pacific Ocean after completing a trip to the Moon. It seems appropriate to us to share with you some of the reflections of the crew as the modern-day Columbia completes its rendezvous with the planet Earth and the same Pacific Ocean tomorrow. First, Mike Collins."

Collins "This trip of ours to the Moon may have looked, to you, simple or easy. I'd like to assure you that has not been the case. The Saturn V rocket which put us into orbit is an incredibly complicated piece of machinery, every piece of which worked flawlessly. This computer up above my head has a 38,000-word vocabulary, each word of which has been very carefully chosen to be of the utmost value to us, the crew. This switch which I have in my hand now has over 300 counterparts in the command module alone, this one single switch design. In addition to that, there are myriads of circuit breakers, levers, rods, and other associated controls. The SPS engine, our large rocket engine on the aft end of our service module, must have performed flawlessly, or we would have been stranded in lunar orbit. The parachutes up above my head must work perfectly tomorrow or we will plummet into the ocean. We have always had confidence that all this equipment will work, and work properly, and we continue to have confidence that it will do so for the remainder of the flight. All this is possible only through the blood, sweat, and tears of a number of people. First, the American workmen who put these pieces of machinery together in the factory. Second, the painstaking work done by the various test teams during the assembly and retest after assembly. And finally, the people at the Manned Spacecraft Center, both in management, in mission planning, in flight control, and last, but not least, in crew training. This operation is somewhat like the periscope of a submarine. All you see is the three of us, but beneath the surface are thousands and thousands of others, and to all those, I would like to say, thank you very much".

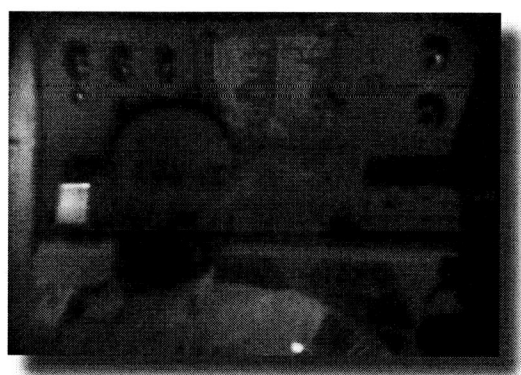

Aldrin: "Good evening. I'd like to discuss with you a few of the more symbolic aspects of the flight of our mission, Apollo 11. As we've been discussing the events that have taken place in the past 2 or 3 days here on board our spacecraft, we've come to the conclusion that this has been far more than three men on a voyage to the Moon; more, still, than the efforts of a government and industry team; more, even, than the efforts of one nation. We feel that this stands as a symbol of the insatiable curiosity of all mankind to explore the unknown. Neil's statement the other day upon first setting foot on the surface of the Moon, "This is a small step for a man, but a great leap for mankind," I believe sums up these feelings very nicely. We accepted the challenge of going to the Moon; the acceptance of this challenge was inevitable. The relative ease with which we carried out our mission, I believe, is a tribute to the timeliness of that acceptance. Today, I feel we're fully capable of accepting expanded roles in the exploration of space. In retrospect, we have all been particularly pleased with the call signs that we very laboriously chose for our spacecraft, Columbia and Eagle. We've been particularly pleased with the emblem of our flight, depicting the U.S. eagle bringing the universal symbol of peace from the Earth, from the planet Earth to the Moon; that symbol being the olive branch. It was our overall crew choice to deposit a replica of this symbol on the Moon. Personally, in reflecting on the events of the past several days, a verse from Psalms comes to mind to me. 'When I consider the heavens, the work of Thy fingers, the moon and the stars which Thou hast ordained what is man that Thou art mindful of him.'"

Armstrong: "The responsibility for this flight lies first with history and with the giants of science who have preceded this effort; next with the American people, who have through their will, indicated their desire; next, to four administrations, and their Congresses, to implementing that will; and then, to the agency and industry teams that built our spacecraft; the Saturn, the Columbia, the Eagle, and the little EMU, the space suit and backpack that was our small spacecraft out on the lunar surface. We would like to give a special thanks to all those Americans who built the spacecraft, who aid the construction, design, the tests, and put their - their hearts and all their abilities into those crafts. To those people, tonight, we give a special thank you, and to all the other people that are listening and watching tonight, God bless you. Good night from Apollo 11."

The camera was then pointed out the window for a shot of the Earth. No doubt, all the engineers who had worked hard over the years since the inception of Apollo to ensure television rode alongside the astronauts were touched by the sentiments of the crew. Although they weren't specifically mentioned, they were undoubtedly a part of the massive team responsible for the success of the mission, and they could justifiably take pride in the cameras they had built which had helped to redefine the way in which humanity viewed itself whether it was from orbit or from the surface of the moon.

A small glitch had arisen due to the last telecast from Apollo 11. Michael Collins explained in the crew debrief on July 31, 1969 , "We made a goof on our last television show. We left the circuit breaker out, which allows the monitor to be operable without transmitting. Consequently, we lost a lot of the entry data. It's the one on 225 called S-band, FM transmitter, data stowage equipment flight bus. Of course, the entry checklist didn't mention checking that circuit breaker, because the people who wrote the entry checklist had no idea that it would be out because of a television program hours prior. I guess the TV checklist doesn't mention it either as best I can recall."

The crew did however, rectify the situation and the splashdown of Apollo 11 occurred in the Pacific Ocean as planned. The USS Hornet recovered the spacecraft once it ended its journey, and a mini TV studio operated by Western Union was on hand to capture the event live. A revolutionary portable TV transmitter built by General Electric, first used for the splashdown of Apollo 7, transmitted the TV signal from ship-based TV cameras to a commercial global satellite, Intelsat III-F-4, 22,300 miles above the Pacific. The signal from Intelsat could then be received by earth stations in Hawaii, Japan, Australia, the Philippines, Thailand, and at Jamesburg, California. From Jamesburg, the signal was sent on via land line to New York where it was than distributed to the major TV networks, to the Manned Spaceflight Center in Houston, and to Etam, West Virginia where it was beamed to another commercial satellite, Intelsat III F-5, over the Atlantic for reception by earth stations in Panama, Puerto Rico, Brazil, Chile, Argentina, Peru and Goonhilly Downs, England. From Goonhilly Downs it was sent to European TV networks via the landlines of the European Broadcast Union.

As the successful lunar mission concluded, scholars of the day prophesized about the rapid pace in which humans would be expected to colonize the moon, and rise to the challenge of landing on Mars and beyond. There was talk of the re-useable "Space Shuttle" which could fly back to earth, and to orbiting space stations. The future looked rosy indeed. But for now, NASA had to concentrate on the remaining lunar missions. When Apollo 11 splashed down in the Pacific Ocean on July 24, 1969 the work may have been over for the returned astronauts, but not so for the people already preparing the next flight to the moon. Those who were involved with TV camera development were also working hard with new breakthroughs expected to be implemented on Apollo 12. The next mission was going to go one step further in the transmission of live pictures from the moon. They were not going to be slow scan low resolution - they were going to be full resolution and full motion - and in color.

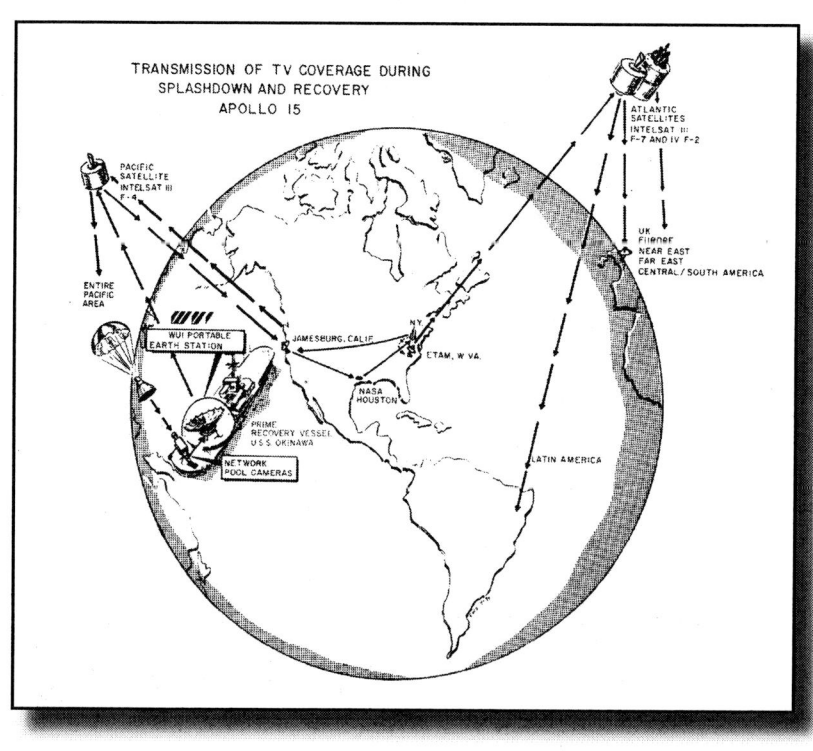

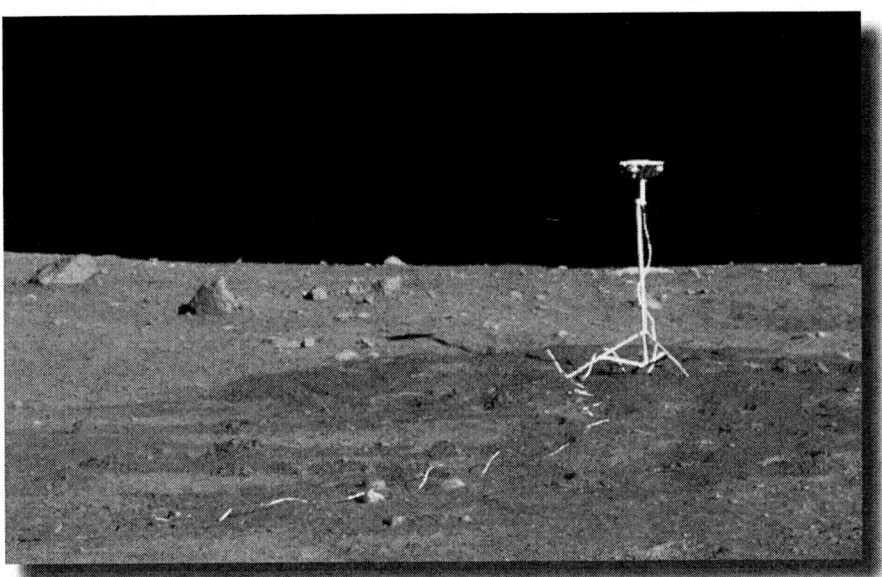

The Westinghouse black-and-white lunar surface camera at the position Neil Armstrong left it on July 20, 1969. It continues to stand there to this day. (NASA Photo AS11-40-5907)

Television Event	GET hh:mm:ss	GMT hh:mm:ss	GMT Date
TV Transmission 1 REC START	010:32:00	0:04:00	17 July 1969
TV Transmission 1 REC END	010:48:00	0:20:00	
TV Transmission 2 START	030:28:00	20:00:00	17 July 1969
TV Transmission 2 END	031:18:00	20:50:00	
TV Transmission 3 START	033:59:00	23:31:00	17 July 1969
TV Transmission 3 END	034:35:00	0:07:00	18 July 1969
TV Transmission 4 START	055:08:00	20:40:00	18 July 1969
TV Transmission 4 END	056:44:00	22:16:00	
TV Transmission 5 START	078:20:00	19:52:00	19 July 1969
TV Transmission 5 END	079:00:00	20:32:00	
Lunar Surface TV (Transmission 6) START	109:22:00	2:54:00	21 July 1969
-TV Lens changed	109:51:35	3:23:35	
-TV Camera moved & Panorama made	109:59:28	3:31:28	
-TV locked into final position	110:02:53	3:34:53	
Lunar Surface TV END	114:25:43	8:01:36	
TV Transmission 7 START	155:36:00	1:08:00	23 July 1969
TV Transmission 7 END	155:54:00	1:26:00	
TV Transmission 8 START	177:10:00	22:42:00	23 July 1969
TV Transmission 8 END	177:13:00	22:45:00	
TV Transmission 9 START	177:32:00	23:04:00	23 July 1969
TV Transmission 9 END	177:44:00	23:16:00	

The television schedule for Apollo 11. The lunar surface portion also includes fundamental changes in camera operations/positioning. The entire broadcast from Tranquility Base was received in slow-scan, low resolution black-and-white.

CHAPTER 11. APOLLO 12

"The thing I remember is that I wasn't worried about pointing it at the Sun. It didn't seem to be a big deal."

Alan Bean.

The mission of Apollo 12, although being the second lunar landing, was also one that contained a number of "firsts". Building upon the achievements of Apollo 11, the mission was to attempt a more scientific assignment, aided by the pinpoint landing at Surveyor crater, the site where the unmanned Surveyor III probe had landed in 1967. Realising the potential for generating huge public interest in the moonwalks, a color TV camera became one of the new things to appear on the mission plans. George Low in a June 26 memo to Sam Phillips announced that there was hurried research into the feasibility of implementing a color television camera into the lunar surface items for the Apollo 12 landing. He mentioned that even from a color camera, the black-and-white image was notably of better quality. He cautioned however, that at such a late date, "… it would be best not to plan on using this system for a September Apollo 12 flight." However, if the flight moved to a November launch date, the camera was ready to fly. With two lunar EVAs planned, the use of a color TV camera was a bold step in showing the world just how the surface of the moon looked, live and in color.

A series of tests were performed on October 19, 1969 to evaluate whether the LM communication could handle the color signal without any hindrance to its communications system. The tests were carried out via the Merritt Island Facility and relayed to the Manned Spaceflight Center in Houston where appraisal was made of the S-Band signal in its lunar surface EVA configuration. The conclusion was not optimistic for the color TV camera. Despite revealing no adverse effects on the entire data package contained in the transmitted S-Band signal, there was notable picture disturbance caused internally by the camera. A pattern of horizontal lines which were a by-product of the color wheel motor and the 100 foot long video cable, which would be connected to the camera as it was activated for the first time, were evident on the images, rendering the quality completely unacceptable. The recommendation was made for a series of filters to minimize the annoying lines, and to assist in better overall signal quality. With the possibility of not having color television beamed from the lunar surface, urgent modifications were recommended to the TV camera to eliminate the problem of the horizontal line interference as quickly as possible.

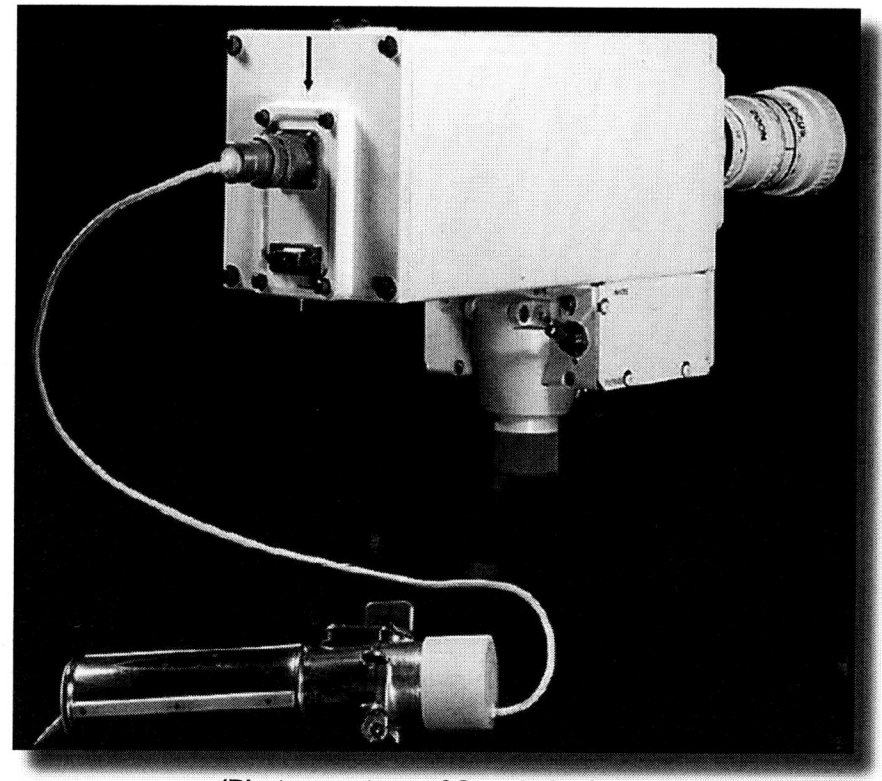

(Photo courtesy of Stan Lebar)

The camera selected for use on the lunar surface was the same one that had flown on the Apollo 10 mission. Modifications were made which were necessary for the camera's thermal distribution requirements while on the lunar surface. These were required because on the airless moon there is no conduction of heat. In direct sunlight the temperature of the camera could get extremely hot, while in the shade it could get extremely cold. If such temperature extremes were to continue for a prolonged period of time, the sensitive electronics inside the camera would be damaged, rendering the camera completely useless. Following the refurbishment, the TV camera sported a new white finish on its casing. The finish allowed as much heat as possible to be emitted from the camera during operation while simultaneously reflecting as much sunlight, thus restricting heat absorption while in direct sunlight. A further modification reduced power consumption of the color wheel mechanism from 20 to 15 Watts and by changing them from metal to plastic, thereby further reducing heat build-up within the camera.

As the launch date rapidly approached, further decisions solidified regarding real time TV support during Apollo 12. According to the proposals at that point, only Goldstone was to be used for television coverage during the flight to the moon. The Honeysuckle and Madrid stations were to only be used for lunar operations. However an energetic response by George Mueller reinforced aspects of the lunar flight which were of great public interest during Apollo 10. Mueller recommended that all three sites be available for TV support for any mission phase that required their TV circuits.

One ambitious proposal for TV use during the moonwalk phase, made by Ed Mason in an October 3, 1969 memo, suggested the crew point the TV camera skywards during their EVA. After zooming in on the planet Mars live for audiences around the world, they could then announce, "This could be the next manned space flight goal during the eighties." Thereby bringing to the attention of all those viewing what the long-term goals of NASA were for the coming decade. This suggestion was unfortunately never followed through. As later lunar missions were flown it was already well established that no such Mars missions would be occurring in the near or distant future.

Apollo 12 launched on November 14, 1969 in the midst of a thunderstorm. In what had become an infamous chain of events, the Saturn V rocket was struck by lightning 36.5 seconds after take-off. Quick thinking advice by John Aaron sitting at his EECOM console had the crew switch the power systems over to auxiliary, thereby forcing the Signal Conditioning Equipment (SCE) systems to revert to the backup mode. The command was a relatively obscure one and for a moment no-one could immediately recall how to implement it. However, thanks to numerous training simulations for the mission, lunar module pilot Alan Bean, spurred by Aaron's suggestion, remembered that the SCE switch was on his side of the control panel. Without wasting any further mission time, he implemented the command and saved the mission from being aborted.

Following the hair-raising launch, and subsequently a full systems check to verify all circuits were operational, the first television from Apollo 12 began during Transposition and Docking. The Lunar Module, still contained inside the SIV-B booster was clearly visible as CM pilot Dick Gordon edged closer and closer to the spacecraft. Once the two craft were joined, the crew began to show astounding views of the earth from the window of the Command module. The Yucatan Peninsula was seen under heavy cloud cover, the same which had seen the Saturn V struck by lightning upon launch. Houston reported back that the earth was indeed round. During the transmission, the camera was switched off periodically despite the TV circuit still being active. Most of the time camera deactivation was done when the crew repositioned the camera from outside views to showing the interior cabin. With amazing clarity Dick Gordon appeared on screen for the remaining portions of the downlink, along with his crewmates.

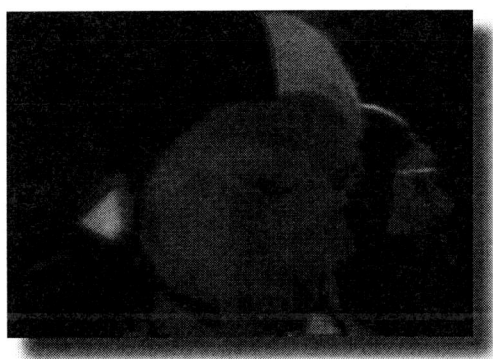

Shortly prior to the end of the telecasts, Pete Conrad enthusiastically recalled the panic of the crew during the lightning strike.

The second transmission was, for the most part, a shot of the Command Module control panel as the crew conducted the hybrid burn for their rendezvous with the moon. Sporting gimmick baseball caps, they were shown carrying out their checklist activities. At one brief point the mini-monitor for the TV camera could be seen.

Following the hybrid burn, a further TV show was conducted inside the LM, where TV audiences could see the stowed PLSS backpacks and the numerous valves required for life support. When the camera was pointed at the earth, the landmass of Australia could be seen with Houston advising the crew that their TV pictures were going out live to the citizens of that country.

"Australia is getting this live? I'd like to say hello to all my friends down there at the tracking station at Carnarvon..."

exclaimed Pete Conrad as the camera shot the earth. Jerry Carr on Capcom gently reminded though, "12, Houston. Don't forget the troops at Honeysuckle." In an effort not to disappoint the technicians working feverishly to receive the incoming TV pictures, Conrad added, "Hello to all the troops at Honeysuckle. Haven't been to Honeysuckle, though." For viewers on the island continent it was a special treat regardless! No doubt further piquing their interest, the crew showed the captive audiences the probe and drogue section of the two spacecraft as the mission flew closer and closer to the moon.

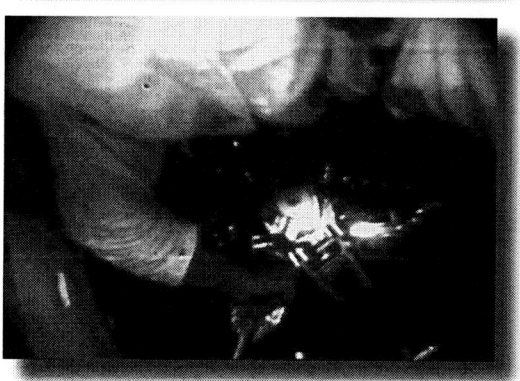

Removal of the probe and drogue after the docking of the LM and CSM.

Just like on the two previous missions, the next TV transmission featured the crater laden surface of the moon in living color. Various landmarks were recognizable, guided by the travelogue given by the crew as they passed over significant surface features. The following telecast featured the slow, yet majestic separation of the Lunar Module on its way to the Surveyor crater landing site. Problems encountered with the direction of the steerable antenna resulted in several moments where the TV signal was lost, yet this did not detract from the interesting views afforded to those watching back on planet earth.

As the egress of Pete Conrad onto the lunar surface began, the Westinghouse TV camera presented the world once again with a view of the moon on the familiar tilted angle due to its mounting in the MESA. This time however, there was a notable clarity from the stunning color image. For Apollo 12 the TV images were not high contrast blacks and whites as seen on the previous mission of Apollo 11, the subtle detail of the ladder could be seen as well as the delicate hue of the moon's surface.

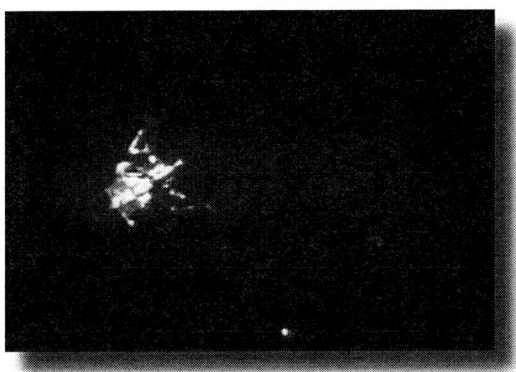

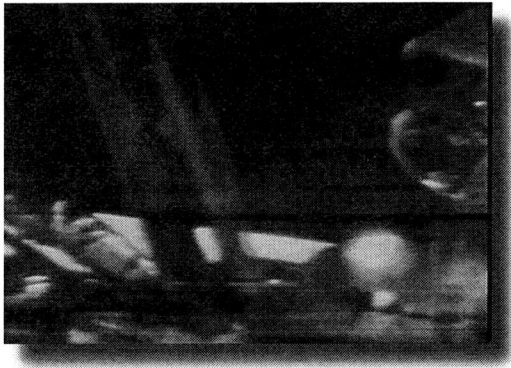

Alan Bean then followed Conrad onto the moon.

Here was a spacecraft about to land on the lunar surface for only the second time in human history. If that wasn't enough to get people excited, the promise of the two anticipated moonwalks in full color certainly was!

Lunar Module Intrepid landed successfully only several hundred meters from the Surveyor spacecraft. Pete Conrad could not contain his excitement as he saw that the spacecraft had flown to its predetermined landing site without fail. He had demonstrated that the guidance system of Apollo was extremely reliable and thanks to it, the two astronauts would relay to the world a spectacular EVA which included visiting the unmanned probe which had landed at the same crater 2½ years earlier.

For almost an hour the world was privy to the moon in color as the astronauts prepared for the EVA with activities near the LM. Every now and then they would appear in shot to access tools and equipment as their checklist EVA plans dictated. Then the moment arrived for the camera to be removed from the MESA to its resting place about 30 meters from the LM, similar to the final position the TV camera had on Apollo 11. In a few short seconds, however, NASA was about to realise how important television was to garnering public interest in lunar exploration.

Without paying attention to the dangers of pointing the camera at extremely bright objects (in this case the sun) Al Bean moved the camera, showing a swaying lunar surface as he did so. No sooner had he swung the camera around on its tripod to point it at the Lunar Module, he also pointed it directly at the bright, unfiltered sun. Instantly the picture became degraded. A huge part of the pickup plate inside the camera tube had been burned into non-existence. Immediately, the automatic level control circuitry kicked in and stopped the iris down; as the missing faceplate had tricked the system into think a bright object was being shot. Everything darkened as the camera electronics attempted to compensate for the bright image. Ed Gibson in Houston reported to the astronauts, "Al, we have a pretty bright image on the TV" although at that point none of them were aware just how serious the brief pointing at the sun had been to the sensitive pickup tube. As Bean tried to find out the cause he received assistance from the ground, "…Could you either move or stop it down?" continued Gibson as the camera continued to display an image that he informed them had "…a very bright image at the top and blacked out for about 80 percent of the bottom."

The first instance of the sun damaging the camera's tube. The damage is not severe and a usable picture could have still been received from the camera.

The second instance of sun damage. The internal circuitry now thinks it is looking at a bright image and stops the camera down. The faint outline of the LM strut can be seen.

In an effort to fix what he thought was a mechanical fault with the color wheel, Bean struck the top of the camera with his geological hammer. Rather than fix the fault, this course of action made things a lot worse. Whatever usable filament inside the pickup tube of the camera remained was removed from its backing, rendering more of the image area useless. Without the backup of the slow scan black-and-white camera, the television coverage from the lunar surface came to sudden and very abrupt end.

As such, there would be no planned television of the first EVA in the near vicinity of the Lunar Module. Much more bitter was that none of the astronauts' examination of the Surveyor 3 unmanned probe would be seen live. What could have been a grand color television event was relegated to an audio-only affair which lacked the immediacy and "being there" aura which would have been evident had the camera not have been damaged.

The full damage following Bean's hitting of the camera with his hammer. The TV camera is now well and truly beyond salvation.

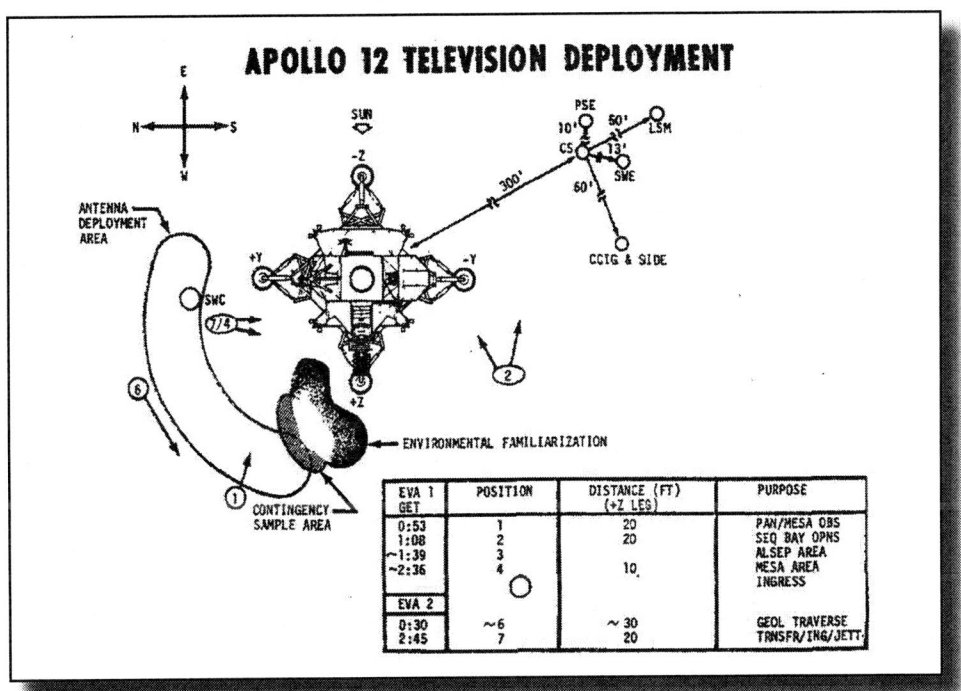

Planned TV Camera Positioning for Lunar Surface Activities on Apollo 12

In post flight debriefings, the astronauts pointed out that during mission training, they only ever had a mock-up of the TV camera. The importance of avoiding shooting directly into sunlight was not given the priority it should have had. Conrad gave his opinion of this information lapse, "Here's what I think the problem was. The TV camera was just like it was part of the comm gear. The TV camera, in that sense, belonged to the comm people...The only experimenters that were going to use anything off the TV were the geologists. But it wasn't their piece of equipment. They were going to get pictures from it. So they told us how they wanted it panned. And remember, the reason they were going to do the pan is they were going to be able to verify where we were. They had some way of analyzing the pictures to say 'Yeah, you're a hundred feet from this crater.' Now, when we did ALSEP and each guy that had a piece of the ALSEP was there, all of that was his gear. But the TV didn't belong to the experimenters (that is, to the geologists). It belonged to operations and they never came out into the field with us."

The ABC network in the USA, faced with no live transmission anymore from the moon, resorted to covering the mission via marionettes to depict the astronaut activities. Other networks had their stock crew of actors doing real time studio simulations of the moonwalks.

However such actions did little to conceal the lack of a truly live picture from the lunar surface. The public was thoroughly disappointed, prompting some viewers to express their concerns directly to NASA.

"I do feel that considering the expense of the equipment they were carrying and the cost of getting them to the moon, they [the astronauts] could have performed their assignments with a bit more discipline and concentration than was displayed in the handling of the television camera." wrote one viewer who relayed the sentiments of many people who watched the drama unfold on their TV sets. He added later in the letter, "…the television failure didn't do a thing for my enthusiasm." Another person reiterated such disappointment by stating, "OK the TV camera is not important to the mission, however more important are the TV transmissions to the people who (again) have paid his [Alan Bean's] way – his apparent unprofessional actions during which he blundered by pointing the camera at the sun is a tremendous disappointment."

Of course the good natured attitude of both Conrad and Bean had nothing to do with the lunar television system's failure. While NASA could not bring to life the broken camera at Surveyor Crater, it was determined to find out precisely what had gone wrong, and so the malfunctioning article was to be brought back to earth for testing and examination--if the fault was one of the camera alone, there would be huge ramifications for Westinghouse. However Stan Lebar was certain that it was only because the camera had been pointed directly at the sun that it was no longer delivering useable pictures. He conveyed his opinion on the matter to NASA by emphatically stating, "...it doesn't mean the camera itself isn't working, it just is the tube wasn't working, and more than likely the camera is still functioning!"

During the scheduled crew press conference while journeying back to earth (using the still intact CSM TV camera) Pete Conrad offered his thoughts onto what had gone wrong with the camera, "Well, Jerry, we really don't know what happened to it. All I know is you told me you were getting a picture and then I didn't pay any more attention to it until I heard you talking with Al, and we don't know what happened to the camera; but we have it on board. We brought it back with us, and whatever is wrong with it, they'll find out and have it fixed so that they have good TV for 13."

Tests involving a similar image sensor to that of the Apollo 12 TV camera were made by exposing the camera system to extreme light. The resulting image as seen on monitors following the replication of the lunar surface mishap revealed that the artificially induced fault was similar to that experienced on the mission.

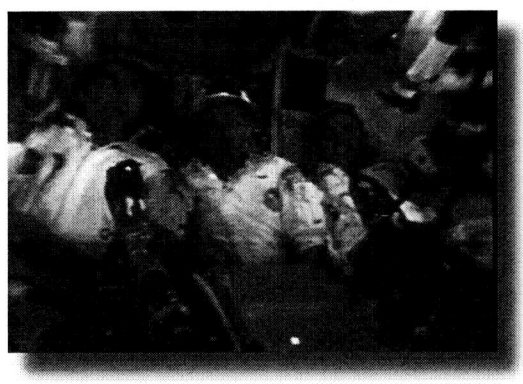

Below: Upon return, the TV camera was taken to Westinghouse laboratories in Elmira, New York where extensive investigation was performed to locate the cause of the fault. (Photos courtesy of Stan Lebar)

Furthermore, after opening a side faceplate to the internal workings of the TV camera, the automatic light-level control was bypassed by cutting one wire. The camera was then pointed on scenery and very good images were obtained in the section of the pickup tube which had been previously rendered black by the ALC circuitry.

Although the damage had been done and television had been lost for all but a small portion of the lunar EVA, numerous suggestions and proposals were subsequently made for improved television system reliability while the camera was operating on the moon. One such suggestion which did not eventuate was the use of the mini-monitor on the lunar surface, given the crews' feedback indicating how hard it was to operate the camera, especially following the tube damage. Additionally, the simple act of supplying a lens cap, and introducing guidelines to ensure the camera was never in a position to be pointed at the sun were also implemented in future mission training and operational procedures. The mishap of Apollo 12's TV camera would never again present itself on any following space mission.

Although the world experienced a great loss in not having the two 4 hour EVAs on television, the mission itself was a complete success, achieving all it had set to do, except of course, full TV coverage on the moon. Compared to Apollo 11, there was a similar amount of television during the flight to and from the moon. Viewers also got to see the Lunar Module approach from the lunar surface and dock with the Command Module; something which was not seen live on the previous mission. When the crew sat together for their in-flight press conference, their camaraderie was infectious. Perhaps the saddest irony is that the mission of Apollo 12 is often overlooked precisely for the fact that the television camera failed one hour into the EVA. Westinghouse and NASA were committed to not having a repeat of the mishap on the next mission, Apollo 13. In a personal response to one of many angry letters regarding the loss of television on the mission, Robert Gilruth promised that nothing would stand in the way of successful television broadcasts from the lunar surface. Not even in his wildest dreams could he have imagined exactly what would not only bring mission coverage to a complete standstill, but which would also hold the very lives of the next three Apollo astronauts on a tenuous thread.

Television Event	GET hh:mm:ss	GMT hh:mm:ss	GMT Date
TV 1 transmission started* TDE	003:25	19:47	14-Nov-69
TV 1 transmission ended	004:28	20:50	
TV 2 Transmission started HYB	030:18	22:40	15-Nov-69
TV 2 Transmission ended.	031:05	23:27	
TV 3 Transmission started.INTERV	062:52	7:14	17-Nov-69
TV 3 Transmission ended.	063:48	8:10	
TV 4 Transmission started MOON	084:00	4:22	18-Nov-69
TV 4 Transmission ended.	084:33	4:55	
TV 5 Transmission started. UNDOCK	107:50:00	4:12	19-Nov-69
TV 5 Transmission ended	108:30:00	4:52	
1st EVA started (egress).	115:10:35	11:32:35	19-Nov-69
1st EVA TV Camera rendered useless due to tube burn.	115:50:35	12:12:35	
TV 7 Transmission started LM ASCENT	145:17:21	17:39:21	20-Nov-69
TV 7 Transmission ended.	145:37:00	17:59:00	
TV 8 Transmission started MOON	172:46:00	21:07	21-Nov-69
TV 8 Transmission ended	173:23:00	21:45	
TV 9 Transmission started PRESS	224:07:00	0:29	24-Nov-69
TV 9 Transmission ended	224:44:00	1:06	

TV transmission times of Apollo 12. Although the EVA transmission was effectively ended upon the burning of the SEC tube TV camera, the camera still was sending a picture. This is evident on viewing the video record in which Pete Conrad can be seen walking in front of the camera. For the purposes of clarity the TX time is listed from TV start to tube burn, where no further TV was feasible. * The TV camera was switched off periodically although the circuit was still up to receive TV. As a result the video record features several dropouts in TV.

CHAPTER 12. APOLLO 13

"Compact color television cameras on the launch pad and on the surface of the Moon promise some never-before-seen views of lift-off and lunar exploration during the mission of Apollo 13."

Westinghouse Press Kit Information, 1970.

The TV camera mishap on Apollo 12 had one positive consequence. It had finally hammered home to NASA just how crucial television was to general public support of the lunar missions. It was only after the numerous outcries from TV viewers who felt they had been robbed of television coverage which their taxes had paid for, that NASA decided to incorporate more coverage into its Apollo flights. Apollo 13 was slated to be the first of a new wave of fully covered missions, from the lift-off, through to splashdown, television would cover the most crucial, and interesting aspects of the operation-- and all of it in full living color!

The first new aspect of daring television coverage would be the launch tower perspective of the Saturn V rocket as it launched into space. Westinghouse was approached by NASA to supply a color video camera to be mounted on the Launch Umbilical Tower. Its sole purpose was to show a perspective of the launch live, which until that time had only been captured by low grade black-and-white TV cameras. For the launch of Apollo 13 TV viewers would be able to see the rocket lift-off in its fiery glory! However, such a proposal was not without numerous problems. The main difficulty being to ensure adequate insulation for the camera as the hot rocket exhaust engulfed the tower.

Working together with the American Broadcasting Company, who operated the TV pool feed from NASA, Westinghouse arranged to have a sequential color camera (the same type as used to transmit color television from the spacecraft) on top of the tower. The view angle would be such that a full "top down" view of the Saturn V could be seen while it sat on the launch pad. The camera was housed in a metal case, shielding it from the expected heat of the rockets as the passed the top of the tower upon launch.

The camera, located at the 360 foot level on the launch tower incorporated a burn-proof image sensor which was designed to withstand the intense brightness of the rocket plume. The new tube largely eliminated the problem of over exposure, which in the case of video cameras could cause lasting damage, as witnessed on Apollo 12. A metal screen mesh was placed over the image receptor which acted as a heat-sink. When a normal video camera is pointed at an extremely bright image the camera focuses this image onto one part of the image sensor and the subsequent temperature increase can occur to as high as 2 million degrees Fahrenheit (1,111,093 degrees Celsius) thus burning the sensor. The mesh would absorb the heat build-up and protect the camera from failure even when images 100,000 times brighter than normal were shot. The launch was to subsequently act as a ground test for the new tube for use on later Apollo flights.

While preparations were underway for the launch camera, refurbishments were being made for the lunar surface camera. While cameras of Apollo 13 would not have the burn-proof tubes, they would have the clever addition of a lens cap which would help ensure that any time the camera was moved, there was no chance that the image sensor could be damaged. Early EVA training footage clearly shows the upgraded camera with the lens cap securely in place during any change of position.

Photographs from this time also show the training MESA complete with the Westinghouse black-and-white slow scan camera mounted in it! Of course, the color camera had always been planned as the lunar surface camera for the mission. Several models of the black-and-white camera were still in existence, and so their use in training procedures was only logical. Why risk damaging a perfectly good color camera when several redundant cameras were available?

By the time the mission actually flew, the black-and-white camera was carried on board as a backup in the unlikely event of failure of the color camera. (NASA Photo S69-57058)

Another little known item which made its debut on the Apollo 13 mission was the very important lens brush. This was to be used for cleaning both the TV and photo camera lenses as lunar dust accumulated on the glass surface during the course of surface activities. Made from camel hair or a similar material it could gently be used to remove the dust as the astronaut saw fit.

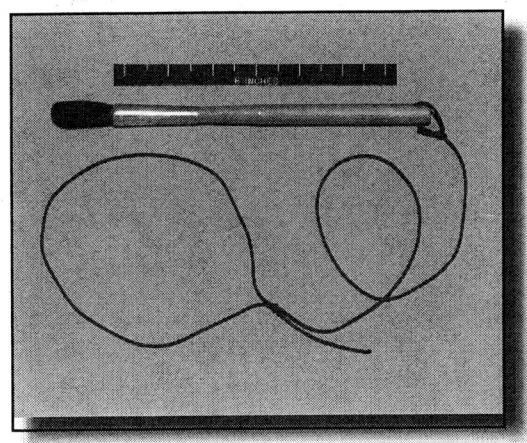

The lens brush which could be used to clean camera lenses on the lunar surface was flown for the first time on Apollo 13 (NASA photo S70-32726).

Apollo 13 launched on April 11, 1970 with the crew of Commander Jim Lovell, Lunar Module Pilot Fred Haise and Command Module Pilot Jack Swigert, who was a last minute change replacement for Ken Mattingly, the original Command Module Pilot, who had been exposed to the measles. NASA could not afford to take the risk that while Lovell and Haise would be on the lunar surface, Mattingly would be breaking out with the illness. The sequential color camera on the launch tower captured the rise of the Saturn V perfectly. The perspective offering a view never before seen live. One could really feel the might of the rocket as it passed the lens of the TV camera.

The fifth engine of the Saturn V shut down prematurely after the launch. An effect known as "pogo" whereby the rocket begins to gyrate as it ascends into the sky causing enough vibration in the rocket to force the engine to stop. However, the other four engines were burned longer to compensate and the launch continued without any further problems.

Just over one hour after launch the first TV transmission began--there had been a brief test of the TV system thirty minutes prior to the actual feed, but this was not recorded. Superb Earth views filled the screen as Apollo 13 flew serenely over the planet.

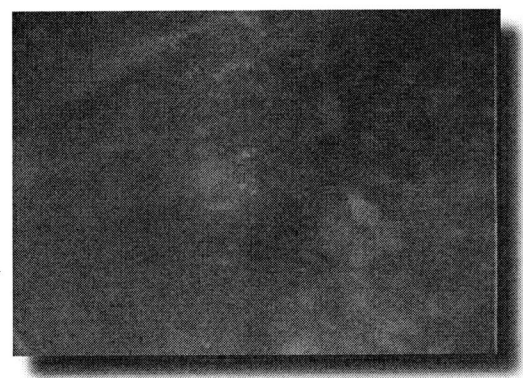

The telecast was very brief, although it did confirm the color system was operating as planned. The next transmission one-and-a-half hours later showed the Transposition and Docking Phase of the flight. Starting with a spectacle of floating debris in front of it, the lunar module, still tightly secured inside the S-IVB booster was clearly visible as Swigert slowly and delicately edged the Command Module closer and closer to the Lander. After several minutes of graceful manoeuvring, the docking occurred flawlessly while Houston watched over the crews' shoulders.

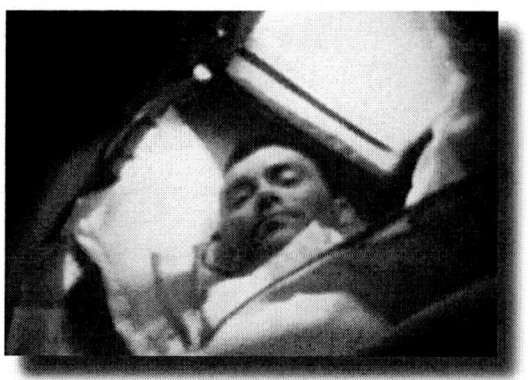

The docking manoeuvre as seen by the Westinghouse color TV camera.

The TV coverage continued as the astronauts inspected the probe connection between the two spacecraft and showed views of planet Earth through the Command Module window. The Lunar Module was then slowly extracted from its cradle and Apollo 13 then began its journey towards its target on the moon. As the crew flew towards their goal, the S-IVB booster could be seen trailing behind them. The abundance of TV coverage prompted Haise to say, "It's been pretty interesting doing all the camera work here because we get a little training running the TV here for when we get on the ground at Fra Mauro. The monitor does make it pretty easy though."

A further transmission occurred during preparations for the Mid-Course Correction burn which was required to place the spacecraft on its heading to properly arrive at the insertion point for the LM to land in the Fra Mauro Highlands. The controls of the Command Module were shown as the astronauts pressed through the checklist ensuring the burn would be carried out correctly. After discussion with ground control, a further TV telecast was brought forward in which the crew of Apollo 13 showed their intended audiences back on earth a guided tour of the Lunar Module. The new red color coding on the helmet which was to be worn by Commander Jim Lovell was displayed where the Navy insignia could be clearly seen.

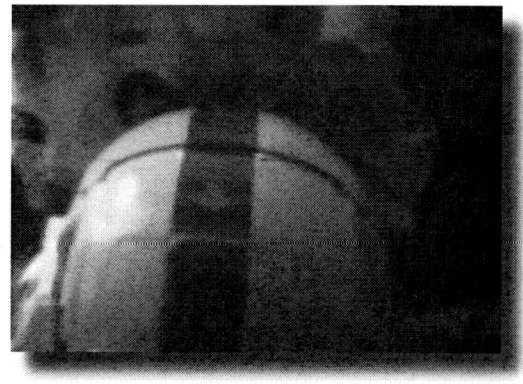

As a practical joke on his crewmates, Fred Haise activated the cabin repressurizing valve which caused a loud bang to be heard in the spacecraft as the air pressure momentarily changed. Lovell commented, "Every time he does that - our hearts jump in our mouth…"

The activities on board showed just how routine a flight to the moon had become, although the TV transmissions were, unbeknownst to the crew, not being picked up by the major US networks. The public had quickly lost interest in lunar voyages, especially as the goal of landing on the moon had been met the year before. Less than an hour after the "guided-tour" transmission, that lack of interest would change dramatically.

After being requested by Houston to stir the hydrogen and oxygen tanks, a procedure performed to ensure the contents had not formed together in problematic layers in the tanks, and also to increase the accuracy of the tanks' quantity readings. Damaged insulation on the wires to the stirrer motor in oxygen tank 2 caused them to short out and ignite the insulation. The resulting fire rapidly increased pressure beyond its nominal limit resulting in the tank exploding. The high gain antenna was struck by debris and communication was disrupted for 1.8 seconds. Swigert responded to the situation by exclaiming, "OK, Houston we've had a problem here!"

Houston completely caught off-guard responded with, "Say again, please." Lovell then reaffirmed Swigert's comments and reiterated, "Ah Houston, we've had a problem."

At first, suspicions were aimed at the TV camera. Stan Lebar explains, "There was a period shortly after the event that the people at mission control thought that the television tube…in the camera had arced causing a catastrophe, since the failure occurred just as the TV transmission had ended. We had gone through a two year period working out the processing details to assure that the high voltage would never arc under any of the conditions from full oxygen through decompression to and recompression to full oxygen. The test data verified that we had solved the problem however, we were all nervous every time the camera was turned on. Evidently, that was the first thing that those at JSC, [who] were familiar with the sensor tube high voltage problem, believed had happened. As far as I knew, the camera power had been turned off and could not have caused the problem. The question that was raised was how long was it from the time they said they were ending the transmission to when they actually turned off the camera power. At that point someone starting yelling that they wanted every bit of data and all data recordings sent to JSC NOW! It took a while but someone came in with the time-line when the event occurred and verified that the camera power was turned off before the problem occurred and the camera could not have caused an explosion. I left at that point and I must admit none of us slept well that night…for a moment our worst fears had been realized."

The teams on the ground then worked around the clock to ensure the situation, though already critical, did not put the crew in further danger. The moon landing was cancelled and the Lunar Module, which had been unaffected by the explosion, became a modified lifeboat entrusted with the task of keeping the astronauts alive until they returned home. Using the Lunar Module's rocket, the free-return trajectory was attained once again, allowing the spacecraft to circle the moon and come back to earth.

In the meantime, the lithium hydroxide canisters which purified the air in the spacecraft had to be jerry-rigged to accommodate the extra person being supported in the Lunar Module as the Command Module was shut down to conserve the all-important remaining power for re-entry into Earth's atmosphere. As two separate contractors had built the two spacecraft, no attention was given to the fact that the canisters were different shapes and incompatible with each other. Instructions were read up to the crew as to how to devise the new units out of material found in the two craft. Thanks to incredible ingenuity on the part of NASA engineers, the crisis of breathable oxygen running out was avoided and Apollo 13 managed to limp all the way back to Earth. No further TV transmissions were made through the remainder of the crippled mission, although Stan Lebar related his concerns to have the public given a direct window into the crisis as it was in full swing by telling NASA, "You need to have enough battery power going for television!" While, of course, such a proposition could not have eventuated given the gravity of the situation, the idea was a daring one. Bringing earthbound audiences directly into the damaged spaceship would have certainly presented television coverage the likes of which would have never been seen before! Given the gradual wane in general interest in lunar missions, such a transmission may very well have drawn attention to NASA's Apollo project in a way no other coverage could.

As it was, the events of Apollo 13 resulted in a morbid curiosity which had changed another routine mission into one which every person on the street was suddenly interested in. This somewhat warped enthusiasm reached the same levels as the interest seen during the successful landing of Apollo 11, and while it was different in nature to that mission, it certainly hammered home to NASA the power of the media and in particular, television.

Had Apollo 13's Lunar Module "Aquarius" landed at Fra Mauro, the TV coverage would have followed the well established format seen on the previous two landings. While still in the MESA, the camera would have shot the astronauts coming down the ladder and stepping onto the lunar surface. After approximately 45 minutes, it would have been removed from the MESA storage location, placed on a tripod and moved to a standing position around 50 feet from the Lunar Module (between the +Y and –Z axes) at an angle of 2 o'clock. The erectable S-Band antenna would have been deployed to allow a better signal to be transmitted from the lunar surface. A panorama would have been scheduled to scan the surrounding terrain, allowing approximately 10 seconds for each field of view.

To avoid any potential problems by pointing the camera directly at the sun, a restriction would have been placed on using the angles within 45 degrees of up-sun. The camera would have also been briefly relocated to show the off-loading of the ALSEP equipment.

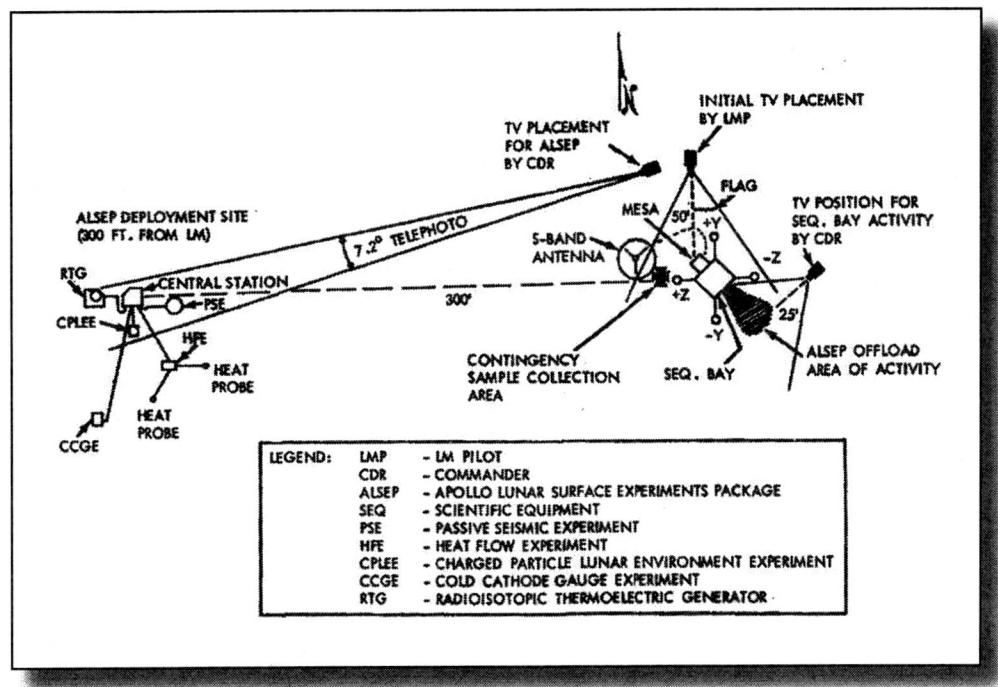

Planned TV placement during the lunar surface activities of Apollo 13.

Following this, it would have then been repositioned to view the ALSEP site. As the camera was limited by the 100ft cable connected directly to the Lunar Module, it would have simply been left to shoot the location from a distance. The astronauts would have appeared as two bright white figures in the distance, but at least they would have been seen in real-time. Once the ALSEP activities would have concluded, the astronauts would have traversed back to the landing site and again turned the TV camera toward the Lunar module. From this position it would have showed the explorers closing out the EVA. Once they had both returned inside the spacecraft, the camera's power would have been switched off until the next day's EVA.

EVA 2 would have begun with views of the Lunar Module and its direct vicinity. Once the astronauts had prepared for the journey to Cone Crater, they would have moved the camera to show the long traverse to the site. There would have been no camera movement at all during this phase of the EVA. Essentially it would have shown a static moonscape for the duration of the traverse once the astronauts had walked far enough away to be no longer visible. Several references to using the telephoto zoom lens are mentioned, although it is unclear how stringent such a recommendation would have been followed had the traverse actually occurred.

Once the astronauts had returned to the Lunar Module after their long trek, the camera would have been repositioned to show the spacecraft. All activities would have been covered from this single vantage point until the astronauts returned into the cabin. Some time prior to lift-off the TV circuits would have been deactivated and the TV coverage would have concluded until the camera in the

Command Module was activated to show the approach of the Lunar Module from the lunar surface. During the journey home, views would have been shown of the receding moon, the approaching Earth and the now-familiar crew question and answer press conference. Unfortunately, none of that ever occurred. The next time anyone on Earth would see the spacecraft or the astronauts would be as it splashed down in the Pacific Ocean.

On April 17, 1970, millions of people watched in anticipation as Mission Control waited anxiously for the familiar sight of opened parachutes. Those chutes would mean the crew had made it safely back to the ground. After the longest re-entry blackout phase of any of the Apollo missions, the TV feed from aboard the recovery ship Iwo Jima showed three successfully deployed chutes as the capsule headed towards splashdown. The crew reported they were in good shape and most definitely in good spirits--happy, no doubt, to be back home. Once again viewers had been deprived of live television from the surface of the moon. While the next mission, Apollo 14 would essentially conduct the exploration that Apollo 13 was meant to, there would be a long wait until any manned spacecraft flew again.

DAY	DATE	CST	GET	Hh:mm	Activity	VEH	STATION
Saturday	11-Apr	14.48	1:35	0:07	Earth	CSM	KSC**
Saturday	11-Apr	16:28	3:15	1:08	TD&D	CSM	GDS
Sunday	12-Apr	19:28	30:15:00	0:30	Spacecraft Interior	CSM	GDS
Monday	13-Apr	23:13	58:00:00	0:30	Interior & IVT to LM	CSM	GDS
Wednesday	15-Apr	13:03	95:50:00	0:15	Frau Mauro	CSM	MAD
Thursday	16-Apr	1:23	108:10:00	3:52	EVA-1	LM	GDS/HSK
Thursday	16-Apr	21:03	127:50:00	6:35	EVA-2	LM	GDS
Friday	17-Apr	9:36	140:23:00	0:12	Docking	CSM	MAD
Saturday	18-Apr	11:23	166:10:00	0:40	Moon	CSM	MAD*
Saturday	18-Apr	13:13	168:00:00	0:25	Moon post TEI	CSM	MAD*
Monday	20-Apr	18:58	221:45:00	0:15	Earth & interior	CSM	GDS
** Tentative							
* Recorded							

The planned TV transmission for Apollo 13.

EVENT	GET	GMT	GMT Date
	hh:mm	hh:mm	
1st TV Transmission START	1:37	20:50	4/11/1970
1st TV Transmission END	1:43	20:56	
2nd TV Transmission START	3:09	22:22	4/11/1970
2nd TV Transmission END	4:20	23:33	
3rd TV Transmission START	30:13:00	1:26	4/13/1970
3rd TV Transmission END	31:02:00	2:15	
4th TV Transmission START	55:14:00	2:27	4/14/1970
4th TV Transmission END	55:46:00	2:59	

The actual TV transmissions made up to the oxygen tank explosion on Apollo 13.

CHAPTER 13. APOLLO 14

"All we did was set up the camera, point it in the direction that Mission Control guys wanted, and adjusted the settings as directed."

Ed Mitchell, Lunar Module Pilot, Apollo 14.

After several months of long and detailed investigations into the cause of the Apollo 13 explosion, Apollo 14 was scheduled to re-attempt the landing, which had been denied to the crew of the previous flight, at Fra Mauro. America's first man in space, Alan Shepard was to command the team consisting of Command Module Pilot Stu Roosa, and Lunar Module Pilot Ed Mitchell. Originally slated to fly the mission of Apollo 13, the crew was pushed back one flight so as to give Shepard the extra training NASA officials felt he needed.

The television system to be flown on the mission was essentially the same as that on Apollo 13. The Command Module color camera was to be used for the cabin interiors and shots of the earth and moon from the spacecraft window during the journey to and from the moon, as had been the case on all flights since Apollo 10. A second identical camera, except for its thermal coating to protect it in the harsh lunar environmental extreme, would be activated as the astronauts egressed onto the lunar surface. It would then cover the two planned EVAs from a fixed location while mounted on a tripod.

The burn-proof image sensor which had been successfully used to transmit live launch tower perspective pictures of the Apollo 13 launch was added to the lunar surface camera. The new tube was Westinghouse's Electron Bombardment Silicon (EBS) sensor. It had the benefit of being able to "see" in extremely low light conditions, yet would not burn out if accidentally pointed at the sun. Due to a wire mesh covering the image sensor, any heat build-up as a result of the sun's highly intense brightness being focused on the senor would be dissipated, leaving the camera undamaged. The ultra sensitive sensor could amplify light ten to twenty times greater than the SEC tube built into the Apollo 11 black-and-white TV camera. As an added precaution, a lens cap was also implemented which would be put over the lens any time the camera was relocated from one location to another.

The slow scan black-and-white camera was also to be carried as a backup for the lunar surface TV camera should it have failed for any other reason. Naturally, there would be no color TV in the wake of such an incident, but at least live images during the moonwalks would be virtually guaranteed. Training film for the moonwalks shows the employment of the lens cap on the mock up camera, highlighting just how serious NASA now took the smooth running of the TV camera. With all these arrangements in camera operations on the lunar surface, there was very little chance of no television occurring during the two planned EVAs.

The Apollo 14 mission, which lasted a total of nine days, launched on January 31, 1971. All systems performed flawlessly and the crew began the manoeuvres for docking with the Lunar Module. TV coverage of the event was started in what was hoped would be a brief operation. However, on the first attempt the two spacecraft failed to properly capture and lock. While the TV images showed the numerous attempts to an increasingly anxious staff of controllers on the earth, Command Module Pilot Stu Roosa concurrently became more and more frustrated. A lot rested on the success of the mission, given the bad publicity brought on by the Apollo 13 mishap. For over 1 hour and 42 minutes, six attempts were made at completing a hard dock, until Mission Control suggested Roosa ram the Command and Lunar Module together. This idea proved successful and the astronauts continued their journey to the Fra Mauro landing site. A television transmission was made in which the probe and drogue elements of the docking section were removed and inspected. The TV camera was used to show Mission Control the exact state of both items in an effort to determine why the docking had been so problematic earlier.

A third TV show along the way featured a guided tour of the two spacecraft. Unfortunately due to the positioning of the Lunar Module it was darker than usual, resulting in very noisy and dim images despite the sensitive camera tube. This did not, however, dampen the enthusiasm of the astronauts in proudly showing off their spacecraft. As a practical joke, the backup crew had placed an abundance of their specially designed and created "backup crew patches" which poked fun at the Apollo 14 crew. Stu Roosa commented on the joy of finding many small "souvenirs" in every conceivable compartment. The cartoon image of the Warner Brothers Road Runner already standing on the moon as a bearded, podgy, red-furred Coyote (a combined caricature of the three men of the prime crew) flew towards it!

After a tense moment where the landing radar did not return any information – requiring a software workaround by MIT to fool the system into accepting the error,

Apollo 14's Lunar Module "Antares" landed in the Fra Mauro Highlands. Once the astronauts had donned their PLSS gear, Commander Al Shepard made his way down the ladder in full view of the color TV camera resting in the deployed MESA. For only the second time ever, color images were arriving live from the moon and this time came the promise that both moonwalks would be presented to the world uninterrupted.

**"Not bad for an old man!"
Shepard on the lunar surface.**

Ed Mitchell follows shortly thereafter.

However, the promise came too late. The American public was already losing interest in lunar missions. Especially after the near disaster of Apollo 13 many were asking whether it was worth the expense and the risk to continue to send up rockets despite the inherent dangers. Incredulously, people wrote in, even going so far as to call their TV networks to protest against interruptions to their viewings of "I Love Lucy" and the like. In today's climate, where NASA has its own TV channel, such emotive responses seem ridiculous and rife with short sightedness. Yet in 1971, it was the status quo.

Al Shepard adjusts the static TV camera with his shadow clearly visible on the lunar surface.

Perhaps the camera never changing perspective unless physically moved by an astronaut had something to do with the public's attention span. As far as they were concerned the astronauts were still doing the same thing they were hours earlier, and the camera was still showing the same view of the Lunar Module. Despite the excitement which should have been present due to a live TV signal from the moon, audiences were bored out of their minds. Baseball telecasts were much more interesting!

Despite public sentiment, Al Shepard and Ed Mitchell had work to do, and they had a very limited amount of time in which to do it. They pressed on with the ALSEP experiment package, leaving the TV camera near the Lunar Module. A minor problem with the camera had presented itself, however. Because the gamma output setting was at 1, it meant no correction was performed in the camera to allow monitors to reproduce an accurate rendition of the scene's brightness. The extreme white of the spacesuits in the direct sunlight caused an effect known as "blooming" any time the exposure was made to yield suitable lunar terrain exposure. Several camera settings were attempted shortly after the TV camera was moved to its location of viewing the Lunar Module and the MESA.

Nevertheless, Shepard and Mitchell moved to the ALSEP site and appeared more like white blobs the further away they traversed. A 16mm Maurer film camera also recorded the work from the MET, so more precise evaluation was possible, though only after the astronauts had returned to Earth.

Despite the blooming evident on the video, the fact was that the live pictures were allowing a window onto the surface of the moon. The immediacy of video, in its full color glory, easily compensated for the dubious exposure problems. Oblivious to the video quality the astronauts continued working, returned to the Lunar Module and after entering the spacecraft, ended television transmissions.

If ever the limited cable of the TV camera was evident, it was on the next day's EVA as the astronauts ventured out to the rim of Cone Crater. Because they could not take the camera with them they simply pointed it in the general direction of the crater and set off without it.

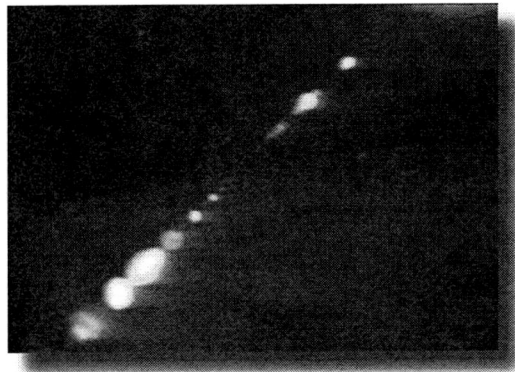

All viewers on earth could see was a few minutes of astronauts moving further away, and the lens flare caused by the sun just out of the TV camera's line of site superimposed over a view of the lunar surface. As voice communications and later photographs noted, the journey was an interesting one. Many diverse rocks were discovered as they moved closer to the rim of Cone crater. Unfortunately, the lack of an atmosphere made judging of distances extremely difficult and so they never made it to the rim. Depth perception was all but non-existent a mere few hundred meters away. Ironically, without realising it the two men had come to within 30 meters of the rim of Cone Crater!

After abandoning the goal of reaching the rim, the astronauts slowly ventured back to the Lunar Module where they commenced the closeout to their second EVA. The cable of the TV camera eventually got caught in the astronauts' boots and it toppled over.

A quick restoration of the camera to its proper position allowed further monitoring of the closeout and also to view a little ceremony Al Shepard had planned. In probably one of the most remembered moments of all the lunar TV footage, Al Shepard hit several golf balls using a specially modified geology tool right in front of the TV camera. Millions of golfers around the world could only dream of being able to hit a few rounds in the 1/6th lunar gravity.

"Miles and miles." Al Shepard plays golf on the moon.

Once everything which was coming back home to Earth was securely packed inside the Lunar Module, the astronauts themselves climbed back inside. After dumping superfluous equipment out of the open hatch, the crew sealed the door and prepared for launch. They then turned off the TV camera circuits, and so ended the 3rd set of transmissions to be made from the surface of the moon.

A minor problem did eventuate while on the lunar surface with regards to the resolution of the TV camera. Towards the second half of the first EVA, this problem began to be very noticeable in the images beamed back to Earth. It was later determined that a temperature increase within the camera caused the focus coil current regulator to overheat, which resulted in the electron beam not being properly focused in the television tube. By keeping the camera operating while in the MESA for 1½ hours prior to its lunar surface deployment, the high-temperature condition was allowed to propagate. The camera was deactivated between the EVA periods to allow it to cool down, despite ground rules stipulating that it remain switched on for this duration. Throughout the second EVA, the resolution of the television picture remained nominal.

TV of the mission returned as Stu Roosa switched on the video camera in the Command Module to capture the incoming Lunar Module. As the lunar surface floated by on the screen, a black dot could be discerned which slowly increased in size. Slowly, the two ships moved together and subsequently docked. Shepard and Mitchell transferred across to the Command module and began their journey home.

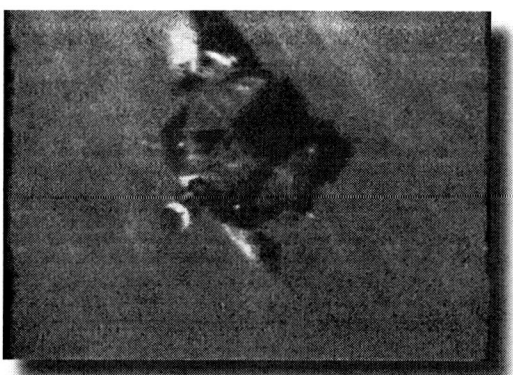

Some time later in the flight a television transmission was made which demonstrated several experiments carried aboard the spacecraft. Viewers on Earth could watch in real-time as the crew demonstrated Heat Flow, Electrophoretic Separation, Transfer of Liquid, and Convection experiments. Of interest was observing the effects of zero gravity on such research.

The Apollo 14 crew conducting their in-flight press conference for the TV camera.

The final in-flight television show featured the crew huddled together for the press conference. Here various questions regarding their lunar mission were asked, mostly discussion over the collected moon rocks, and the nearly successful trek to Cone Crater. The one major point highlighted by the mission was the extremely limiting nature of exploration which could be achieved on foot. The constraint of time was what kept the lunar surface activities from being fully exploited. The introduction of the Lunar Roving Vehicle (LRV) would be instrumental in providing a much more detailed examination of the surrounding area. One important factor remained. If television was to be made of the missions using the rover, which could drive several kilometres away from the Lunar Module, it would be near useless if the camera was left at the LM. A method was necessary to provide live television from each geological site which would adequately cover the astronauts' activities, while not presenting crucial time in its operation.

The Apollo 14 Mission Report published in May of 1971, recommended, "A remotely operated camera with adjustment of focus, zoom, and lens setting controlled from the ground would be very useful in making available lunar surface time presently required for these tasks." The new camera would have to be an astounding piece of engineering, which, like most things in the Apollo timeline, gained from the experience provided from earlier missions. What was to accompany the astronauts on Apollo 15 to the lunar surface would be just such a remotely controlled color TV camera which would be mounted on the Lunar Rover.

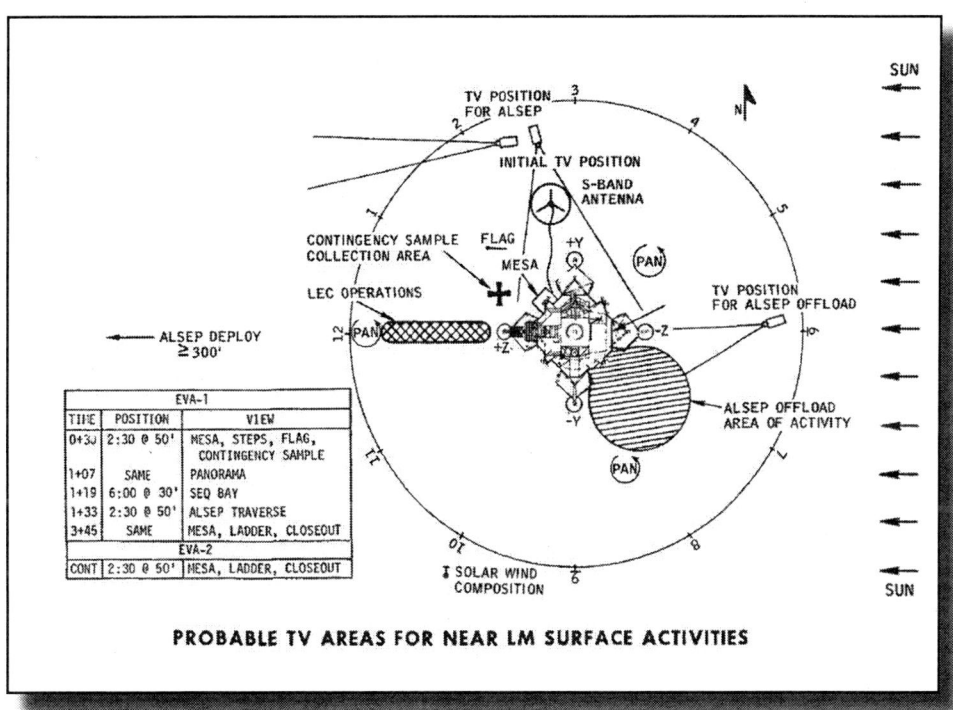

Apollo 14 planned activity near the LM.

Television Event	GET Hh:mm	GMT HH:mm	GMT Date
TV Transmission 1 START TD&E	3:05	0:08	February 1, 1971
TV Transmission 1 END	5:00	2:03	
TV Transmission 2 START PROBE	11:00	8:03	February 1, 1971
TV Transmission 2 END	12:12	9:15	
TV Transmission 3 START	60:05:00	9:08	February 3, 1971
TV Transmission 3 END	60:42:00	9:45	
EVA 1 TV Transmission START	111:14:00	14:43	February 5, 1971
- Panorama (LMP)	114:50:00	15:53	
- TV Transfer to Equipment Bay (CDR)	115:05:00	16:08	
- TV positioning (CDR)	115:22:00	16:25	
EVA 1 TV Transmssion END	118:20:00	19:23	
EVA 2 TV Transmission START	131:40:00	8:43	February 6, 1971
EVA 2 TV Transmission END	135:42:00	12:45	
TV Transmission 6 START	143:15:00	20:18	February 6, 1971
TV Transmission 6 END	143:20:00	20:23	
TV Transmission 7 START	143:28:00	20:31	February 6, 1971
TV Transmission 7 END	143:35:00	20:35	
TV Transmission 8 START	171:30:00	0:33	February 8, 1971
TV Transmission 8 END	172:20:00	1:23	
TV Transmission 9 START	194:29:00	23:32	February 8, 1971
TV Transmission 9 END	194:52:00	23:55	

TV Transmission Schedule from Apollo 14

CHAPTER 14. APOLLO 15.

"You have to understand, we were not television, or camera people. We were assigned to do this job and we went off to do it...the next thing I knew my name was known around the world..."

Ed Fendell, remote operator of the RCA GCTA TV camera on the moon.

The idea of a lunar mission using a vehicle to travel longer traverses was long part of lunar mission planning at NASA. The problem that confronted mission planners was how to make the vehicle compact enough to fit on the Lunar Module. A joint design by Boeing and General Motors realised this practical impasse by allowing the lunar rover to be modular in design. To further facilitate the missions an idea was suggested which would once again push the envelope of existing television technology. As the rover would travel significant distances from the LM, the 100 foot TV cable would not be sufficient to transmit signals for the distant exploration sites.

When the "J" missions, that is those which would use a vehicle to travel further from the Lunar Module, were being designed by engineers at NASA, William E. Perry at the Engineering and Design Directorate recommended that a remote controlled TV camera would be desirable for the missions. Ed Fendell, who would later operate the TV camera from his INCO console in Houston for the Apollo 15, 16 and 17 missions explains, "Bill was an engineer over in the Telecommunications Division of engineering at JSC. He came up with the idea of the capability of controlling that camera." It would be a company which already had TV camera experience for the Apollo missions that would try and regain ground it had lost to competitor Westinghouse during the development of the lunar surface camera.

RCA had presented NASA with several unsolicited proposals for TV cameras on the Apollo missions. The company had assembled a task force from Lancaster, Pennsylvania and Princeton, New Jersey (along with other RCA plants) with the goal of getting RCA more involved with color television from outer space. A prototype was made with which it was hoped, NASA would reconsider the contracted supplier of color TV cameras for Apollo 12.

The demonstration took place on June 11, 1969 in Building 15 to a group of NASA representatives. Although advised that no funding or requirement existed for additional camera development for Apollo 12, RCA emphatically stated that their research would proceed regardless of NASA funding or not. RCA wanted its name closely associated with lunar exploration and it was committed to getting its lunar camera on the flight of Apollo 12, regardless of cost.

A meeting was held on October 29, 1969 to discuss the proposal of a portable TV system considered for the "J" mission where the Lunar Roving Vehicle was to be used. The desire of the geologists for a full resolution system is evident in a follow-up memo from W.E Stoney on November 6, 1969 where he states, "It is the opinion of this program office that this requirement for increased resolution is a valid one and should be seriously considered." Based on these discussions, the investigation into slow scan versions of the TV camera did not appear to progress beyond anything but comparison tables attached to the memos discussing the TV requirements. What is interesting is a request for a transmitter to be part of the TV camera system when direct communications was not possible on the long traverses.

Bellcomm presented a report which outlined the various situations expected on the moon and their associated camera requirements. Briefly, the idea of a high resolution slow scan camera was bandied about and endorsed by the scientific community, although already in the early consideration phases, preference was given by Bellcomm to a TV system equal to network television quality with a full or near full frame rate of 30 frames-per-second. A requirement for the higher frame rates was the availability of 210ft receiving antennas at the ground stations for good reception of the signal.

A meeting was held on December 17, 1969 in the office of Jim McDivitt to discuss the remote control capability of the television system. Robert Gardiner was given the task to determine the mast lengths and azimuth-elevation mounts for the camera and to report his findings to the Flight Operations Directorate at NASA. He was also to report on the earliest date for the availability of a working prototype of such a TV system. The impact of modifications to provide real time command interface between the Mission Control Room and the LM Digital Uplink Assembly was to be handled by S. Sjoberg, while O. Morris and Owen Maynard were to investigate logistics for the mounting of the camera onto the Lunar Module, and to also determine the field of view from various anticipated levels above the lunar surface.

Originally schedule as an "H" mission similar in EVA structure to Apollo 14, where the astronauts were mobile only on foot and thus greatly limited in their traverses, the flight of Apollo 15 was extensively modified on September 2, 1970 after NASA revealed that it was cancelling what were to be the then-current incarnations of the Apollo 15 and the Apollo 19 missions. To take full advantage of the remaining flights, Apollo 15 was redesignated as a "J" mission, which meant it would be the first mission with the Lunar Roving Vehicle. The earlier documentation refers to the Apollo 16 mission, as at that time it was to be the first of the missions to utilize the lunar rover, and also the new remote TV camera system. Once Apollo 15 was modified to incorporate the rover, it became the mission referred to in the numerous studies and design considerations of the new camera.

An illustration depicting the planned communications relay from the LRV.

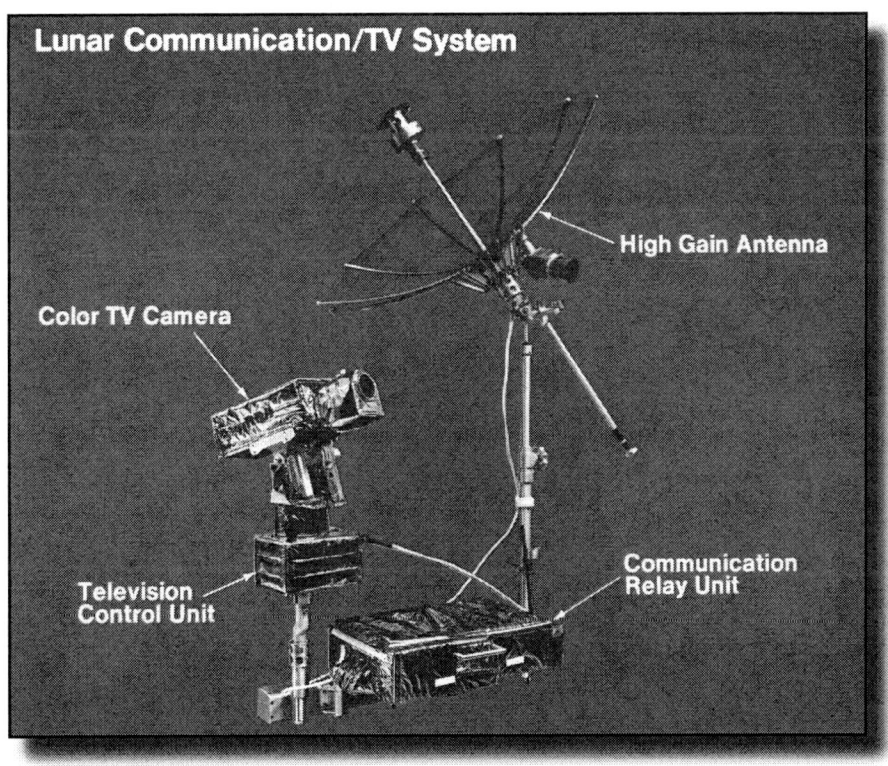

Contract NAS-11260 was awarded to RCA on July 31, 1970 for the procurement of the TV system to be used on the Lunar Roving Vehicle. RCA itself was informed of the decision officially on August 4, 1970 and the camera to be developed was formally known as the RCA Ground Commanded Television Assembly (GCTA) and would be the sole camera used on the lunar surface for the transmission of live television pictures back to earth.

The television camera itself was part of a combined communications subsystem of the Lunar Rover. The Lunar Communications Relay Unit (LCRU) handled the routing of communication signals from the LRV to the earth. The device utilized an uplink dish made from wire mesh which sat at the front of the LRV. The dish was a high gain antenna with an integrated bore-sight to help the astronauts align it with the earth. While the rover was driving it proved near impossible to maintain a fixed view of the earth, so voice communications were routed through the astronauts' PLSS antennas through the LCRU to the low gain antenna. When they had parked the Rover, the astronauts would then quickly align the high gain antenna and switch voice and video communications through the LCRU to be uplinked to earth.

Testing of the LCRU system began as early as November 26, 1969 and used a black-and-white camera. Further comparisons were made with slow scan TV cameras and the Command Module color TV cameras. One clear result of the tests was the severely limited resolution of the picture when the expected signal strengths from the lunar surface were used. Because of this, the immediate recommendation was for a slow-scan system to be used in order to have better resolution and a better signal-to-noise ratio, unless image quality could be increased through further research.

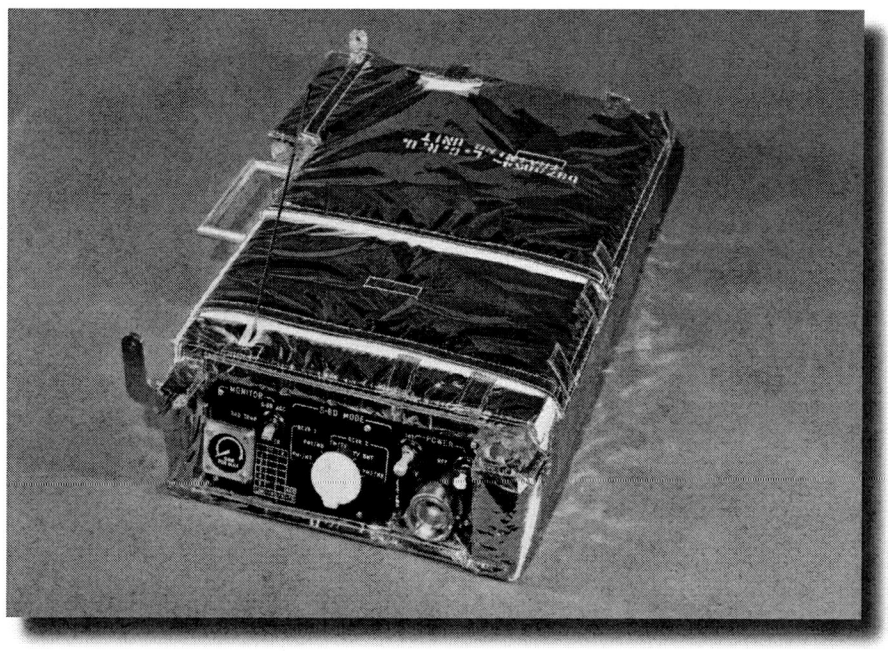

The LCRU unit with thermal blankets fitted. Photo courtesy Sam Russell.

Further tests were conducted which allowed the research teams to focus on a camera which could answer the needs of resolution, color reproduction and size and weight considerations as dictated by the parameters set by NASA. Rocco Petrone outlined his desire in a memo dated December 23, 1969 that, "...the tests will answer as many of the operational, scientific and functional questions concerning the relative camera capabilities as possible."

Pre-emphasis and de-emphasis tests (a type of noise reduction control when used together) revealed surprisingly good overall improvements to the color television pictures through the LCRU. At the uplink site, the TV signal was passed through circuitry that attenuates only the lower frequencies which had the effect of boosting the high frequency portion of the signal. At the receiver site, a de-emphasis circuit attenuates the upper frequencies to restore the original video waveform. The purpose of this technique was to improve the signal-to-noise ratio by about 2 dB which was inherent in the distances travelled by the signal. These tests were performed between April 27 and May 7, 1970 with the recommendation that pre/de emphasis be used for the television transmissions from the moon. Some synchronisation problems were encountered when attempting to matrix the sequential color sequence, but these were resolved with the introduction of processing amplifiers prior to the emphasis application.

Results of pre-emphasis and de-emphasis tests of the RCA GCTA TV system. (NASA Photos S70-26229, S70-26228, S70-26251 and S70-26258)

It was however later testing with the improved model of the GCTA camera where the use of de-emphasis was strongly recommended to be not used. The improvements on the camera yielded much better results than had been experienced on all earlier tests and it was determined the pre/de emphasis use actually degraded the image. Additional testing of remote operational transmission were lost or received as spurious commands in numerous test configurations.

Early in September of 1970, a preliminary design and system interface review was held by RCA. Recommendations were made to incorporate a thermal sensor on the Vidicon faceplate to provide an indication of faceplate temperature. Further thermal and de-emphasis considerations were recommended to be given further tests over a 20 month period. In October senior RCA engineers reviewed the design of the camera and with additional systems management support, worked to improve the overall video quality of the camera.

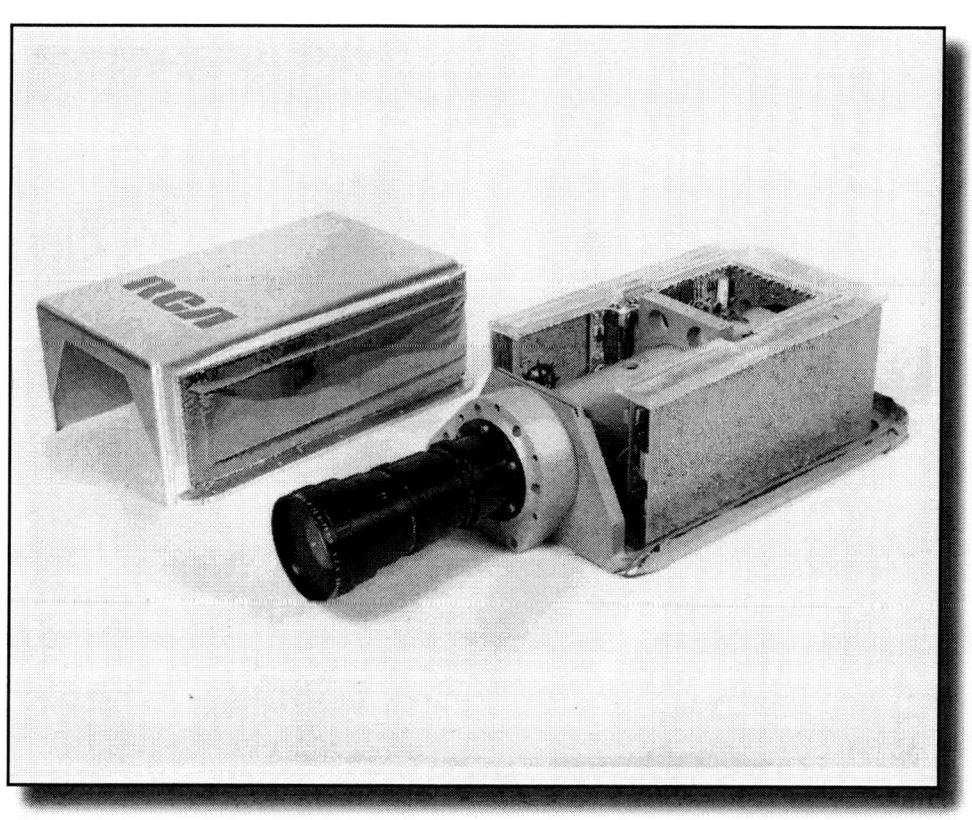

**The RCA GCTA TV camera showing the electronics inside the camera housing.
(RCA Astro Electronics Division Photo 69-11-60C)**

(RCA Astro Electronics Division Photo 72-11-532)

The in-house effort devoted to developing and testing the camera became known as the "Rainbow" team. It was responsible for ensuring the cameras fully complied with NASA's mission parameters. Two GCTA models were delivered to MSC for evaluation. While originally seven flight models were planned to be supplied to NASA, on October 28, RCA was notified by the agency that two models were to be deleted from the contract, leaving five deliverable Flight Units. In order to prolong the shelf life of the cameras, it was suggested that they be turned on every 90 days for a total of 30 minutes.

Westinghouse, somewhat shaken by RCA's encroachment into the television portions of the mission, fiendishly devised their own team's name for color TV development. "FURCAT" was a name made by initializing a sentence which cheekily described the attitude of WEC toward RCA! Using one's imagination will quickly reveal the meaning of the letters.

Dust reliability tests were extensively performed on the GCTA. It was paramount that the camera still perform acceptably, even when covered in lunar dust and thankfully the results were extremely positive that the system could withstand the abrasive lunar dust even when significantly exposed to it.

Thermal vacuum testing and vibration tests as well as acceleration tests were all performed in order to replicate mission situations as accurately and as thoroughly as possible. As Sam Russell explains in "Lunar and Planetary Rovers: The Wheels of Apollo and the Quest for Mars", by Anthony Young, "We had a huge thermal vacuum chamber in East Windsor, New Jersey. We also had to test every piece of equipment in a 100 percent oxygen environment, regardless of whether it was inside or outside the Lunar module, as a result of the Apollo 1 fire." Additionally, Russell created simulated lunar scenery to assist with the color tests of the camera. These were essential in determining how dark the sand would appear on the television image, and thus facilitated in the camera's sensitivity development.

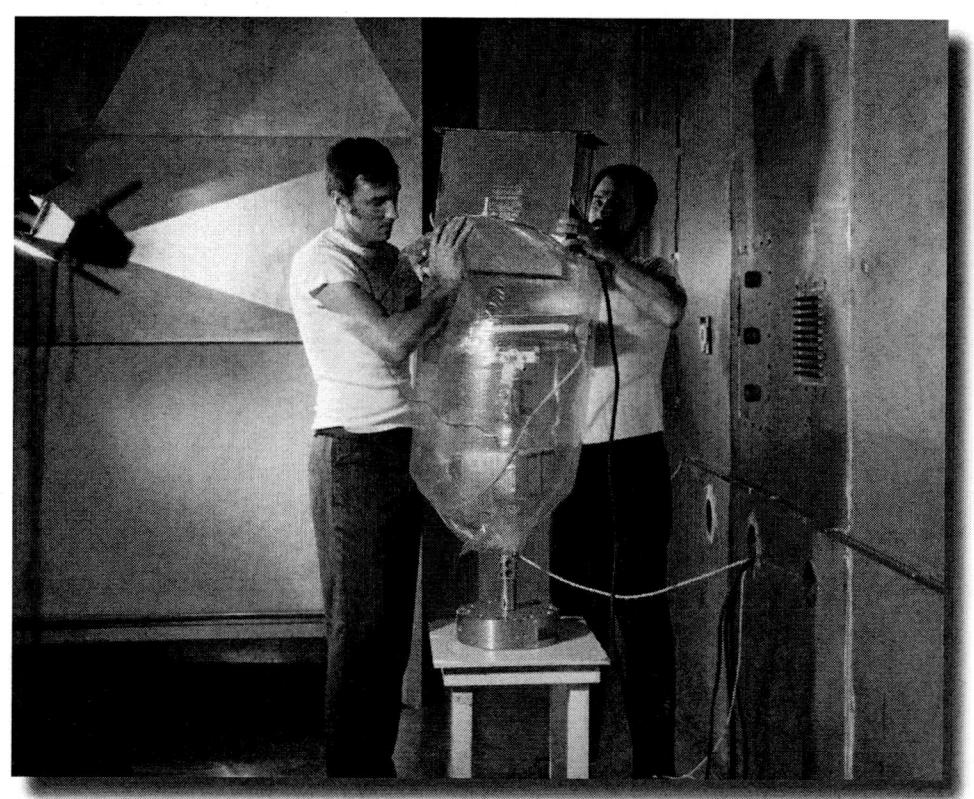

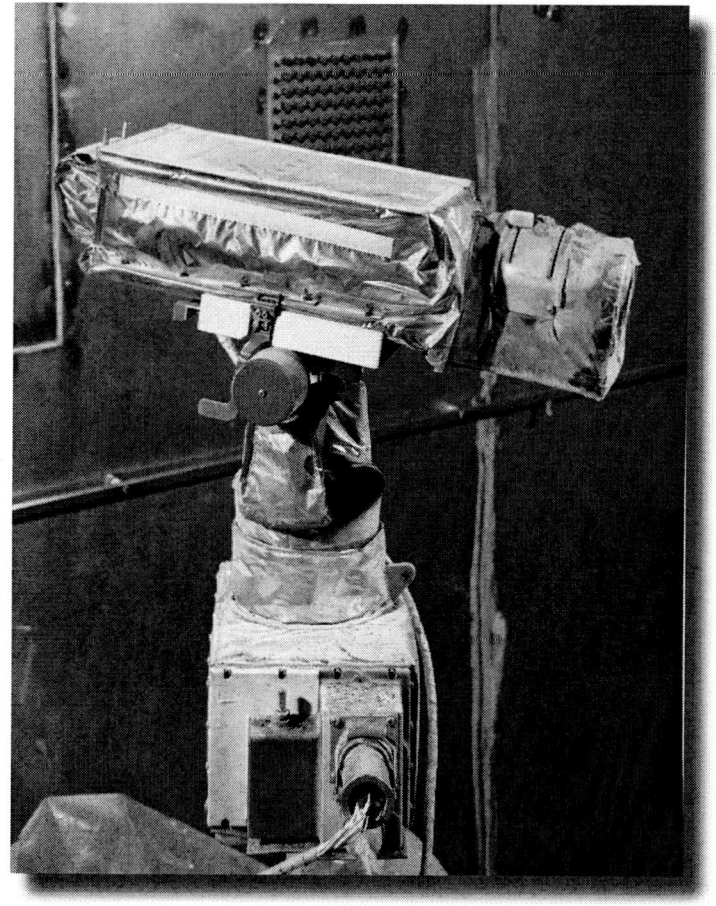

Previous page: Dust testing on GCTA Camera. This page: Model lunar surfaces created by Sam Russell for TV camera testing. Photos courtesy of Sam Russell.

Testing of the internal circuitry and mechanical functions of the TV system were conducted through to the end of 1970. Thermal considerations continued to be the main focus of physical design attributes of the camera. It was determined that thermal blankets covering the camera would be required from the first EVA. Automatic Light Control and Automatic Gain Control functions were added to the system to facilitate in an improved TV picture. According to Sam Russell, the picture tube had to undergo tests where it was pointed at the sun for extended periods of time and not be burned out because of it. NASA was adamant that a repeat of Apollo 12's TV problems would not occur with the GCTA.

A recommendation was made by Bellcomm to reduce the thermal radiation requirements of the LCRU. A drop in power from 10 Watts to 5 Watts was proposed with either the size of the parabolic dish staying the same 30 inches, or by increasing it to 42 inches.

One factor of camera control which had to be fully addressed was the matter of the incredible expanse between the moon and the earth which the commands for the camera would have to travel. Approximately 2.5 seconds were required for the signal alone to travel to the moon and back. Additionally delay factors inherent in the TV system introduced a further 0.5 seconds of delay if only the Goldstone tracking station was used in the transmission/reception of the command signal. When the Australian stations (Parkes or Honeysuckle) were used, there was an additional 0.6 seconds of delay introduced by the satellite network sending the signal back to Houston. The controls for the camera included panning left or right, tilting up or down, zooming in or out, and control the aperture for correct exposure.

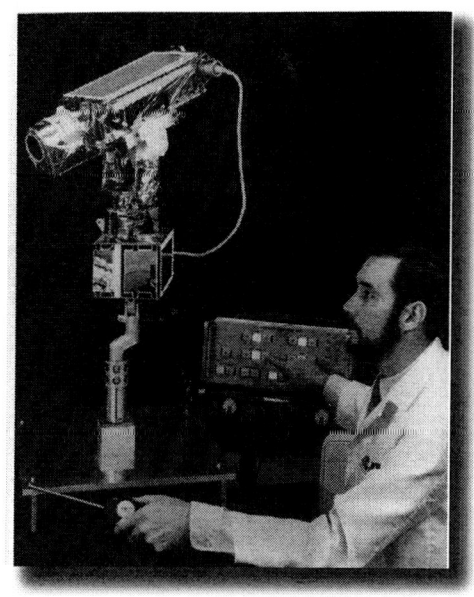

Lens tests (top of page) and testing of the remote control functions of the RCA color camera (above). (RCA Astro Electronics Division Photos 70-5-524 and 72-2-521)

The Bellcom report details the command system as follows:

"Television camera commands will be in the form of Real-Time Commands (RTC's). RTC's are stored pre-mission in the core memory of the command computers at the remote MSFN sites. They are called up during a mission by Execute Command Requests (ECR's) initiated by Flight Controllers at the Mission Control Center (MCC). Upon receipt of an ECR, the requested RTC is called out of memory and a "vehicle" address (in this case, TV camera) and "system" address are added. The entire message is then "sub-bit" encoded (each information bit is transformed into a 5-bit code for error control purposes) and uplinked at a 1-kbps rate on a 70-kHz subcarrier of the Unified S-Band System. The uplink composite signal will be received by the Lunar Communications Relay Unit. (Remote control of the camera through the LM systems is also being considered.) The 70-kHz subcarrier will be sent to the TV command decoder, which will demodulate the signal, check the address and sub-bit structure and, if correct, activate the appropriate relays for television camera control. Unlike other commands, no Message Acceptance Pulses (MAP'S) will be returned to the Earth from the LCRU. This will require either that the MSFN site uplinking the command be in a "MAP Override" mode or that the Flight Controller(s) at MCC concerned with the TV control ignore the REJECT signal they will otherwise receive from the MSFN site in the absence of a MAP. The TV command system is not a closed-loop system in the same sense the other Apollo space vehicle command systems now are. The observed response of the television camera will close the loop, not the receipt of a valid command indication. Consequently malfunction diagnosis may be more difficult."

Taking the delay times into consideration, the suggestion was made for extensive training using the expected delays in order to enable ground control staff to become accustomed to the delayed response. Such training occurred by way of simulated lunar environment and mission practice runs. Speed times for pan and tilt were determined to be optimal at 3° per second. Naturally, when the zoom lens was used this speed would appear faster on features further from the Lunar Module as the zoom feature would magnify any movement of the camera. However, in general it was felt this ratio allowed reasonable control given the intervals expected.

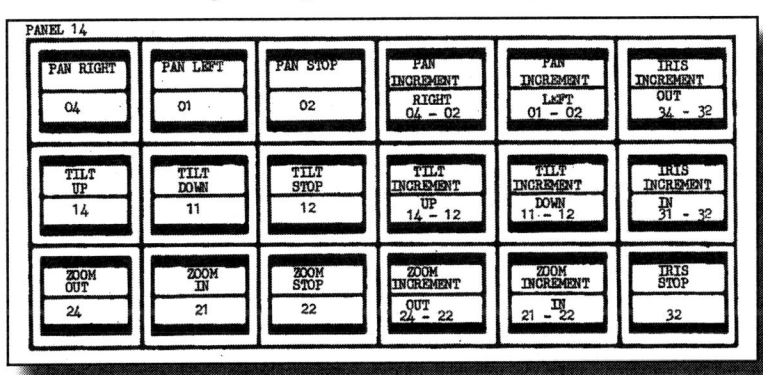

The panel layout used by Ed Fendell to remotely control the GCTA from Houston.

There had been early discussion on the feasibility of continuously up-linking video during traverses by use of a medium gain antenna similar to that used on the ALSEP experiments. While more robust than the high gain antenna method, it too was severely limited by terrain, having a margin of only 8° variation in slope before signal would be unusable. Testing quickly determined that such a configuration was not feasible, as keeping a good line-of-sight with the Earth while driving on the undulating surface of the moon was near impossible. During the later missions there would be several brief moments of active television while the rover was in motion. It is evident how easily the signal could be lost by even the most minor of movements during a traverse.

A suggestion extending from the idea of continuous television coverage during traverse was made by Dr Farouq El-Baz as early as 1968. This involved remotely driving the rover in between missions from one landing site to the next, and in the proposal given by El-Baz, this was hoped to be the Hadley Apennine region. The television system was considered in a function of being the "eyes" of controllers steering the Rover from Earth. The unmanned Rover would have landed 500km from the Hadley site six months prior to the manned mission. The Rover would then drive a predetermined route to the future landing site remotely gathering samples from the diverse regions driven through. Dr El-Baz explains, "The reason for my idea was to be able to photograph and sample the interior of the Hadley Rille. Although most of us assumed that it was the product of a lava tube, there were those that invoked water erosion. Thus, my idea to resolve that and obtain information from one of these peculiar lunar features. NASA was not moved because of the cost and trouble for a mission that had only scientific objectives and not much engineering…" The question is still open as to whether the TV camera could have been feasibly activated to cover the landing of the approaching manned Lunar Module had such a mission been given the go ahead. Such an event broadcast live on the TV networks would have been absolutely spectacular!

As the camera was fully operational via remote control, and powered by the Lunar Rover batteries, the camera could stay operational for some time after the Lunar Module had left the lunar surface. Survey of the landing site was prepared in the mission plans in addition to an idea which would potentially be of great interest to scientists – observation of a solar eclipse made as the Earth passed in front of the sun.

In the months leading up to the launch of Apollo 15, numerous engineering examinations were performed on the GCTA system by NASA engineers. Richard Bohlmann kept meticulous notes about the various tests in a notebook which was kept by his nephew, Dave Bohlmann, who relates, "It looks to be a lab book concerning final integration testing of the LRV's Ground Controlled Television Assembly (GCTA) and the Lunar Communications Relay Unit (LCRU)…one entry says 'data drop-out during rover running at 10 km/hr'. It's mainly about crystal oscillators, transmitters and computer commands - things that did not work per spec, of course." The entries as documented related to the methodical assessment of discrepancies in the unit prior to flight. Many instances related to problems encountered prior to mission simulations, in which a practice run of the fully operation TV system was put through its paces, as well as flight units which were to be used on actual lunar missions.

Excerpts from the engineering notebook detailing tests on the control system. Scans courtesy of Dave Bohlmann.

The model of the GCTA TV Camera to be used on Apollo 15 was delivered to the Manned Spacecraft Center on April 15, 1971 for fit tests and final electrical testing. On April 26 it was returned to the RCA Astro Electrical Division for incorporation of any final changes and for completion of acceptance tests. By May all relevant testing and acceptance had been completed and on May 21 the camera was delivered to MSC for inclusion into the Lunar Rover Package destined for the moon.

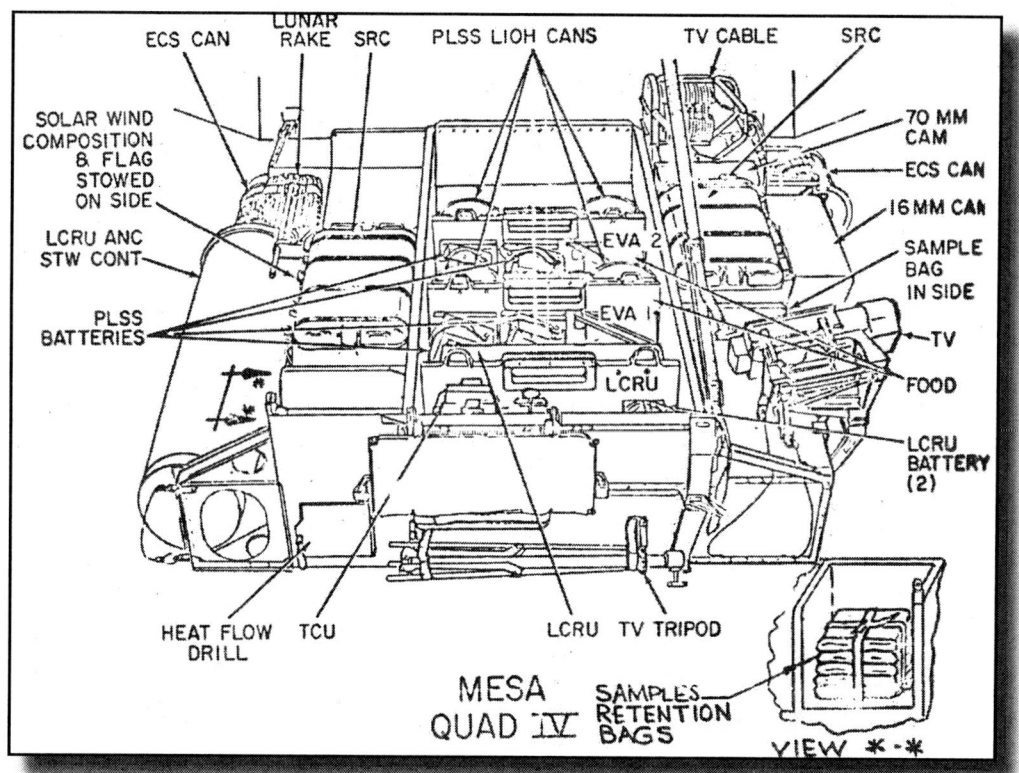

For the first and only time on the remaining lunar missions, the RCA camera in its stowed configuration in the MESA televised the descent onto the lunar surface by both Dave Scott and Jim Irwin. (RCA Astro Electronics Division Photo 75-3951)

Apollo 15 launched on July 26, 1971 with an all-Air force crew of Commander Dave Scott, Command Module Pilot Al Worden and Lunar Module Pilot Jim Irwin. Transposition and Docking were broadcast live using the Command Module television color camera. While RCA had provided the lunar surface camera for use on the rover, the Westinghouse camera was used for all in-spacecraft television throughout the flight.

Following the flight to the moon followed by a landing which seemed to be getting more and more routine, the astronauts prepared for three days of activity on the lunar surface. One important step made on Apollo 15 (which did not occur again on the remaining missions) was a stand up EVA made shortly after landing. Dave Scott surveyed the area and described surface features to the geologists. The survey data greatly aided the on-the-fly design of the explorations which then occurred at the landing site.

Both he and Jim Irwin proceeded to unload the vehicle under the watchful eye of Mission Control back in Houston. At that point the camera was operating just like all the lunar surface TV cameras before it, namely static and in a fixed position on a tripod. For the entire time that the crew configured the Rover for its lunar environment the camera remained in this position, moving only when an astronaut adjusted it.

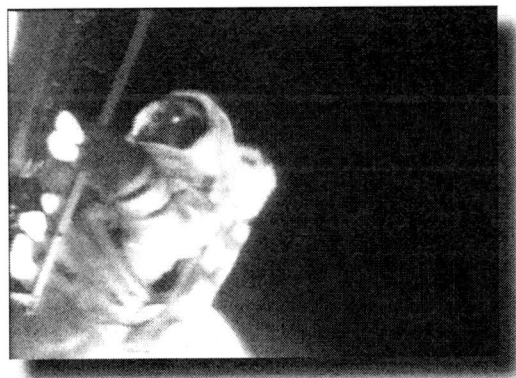

Once they had completed their preliminary tasks around the Lunar Module, Dave Scott moved the camera to an area still in the shade of the spacecraft and pointed it at the section which stored the Lunar Rover.

However, the best was yet to come. Gradually the combined communications package was assembled, requiring only the TV camera to be connected. Dave Scott then approached the GCTA camera on its tripod and moved it to the Lunar Module. He then advised Houston that picture would be momentarily lost as he connected it to the fitting on the Rover. All those who had worked hard to bring the reality of a remote controlled TV camera on the moon into being, collectively held their breath. Now was the moment of truth. Scott reported to Houston, "Okay; going to TV Remote." This meant that the camera had been taken out of local control and was now able to receive commands from Earth. He then said to Houston, "Okay; she's all yours." At first there was no activity or picture to be seen from the camera. Dave Scott verified the camera control switch one more time and as he confirmed his actions to Houston, the GCTA TV camera sprang to life.

"Presto chango; there's the TV!" came the response from Joe Allen in Houston as the camera sprang to life on its new fixture the Lunar Rover. Back on earth, TV audiences had grown to be the largest since the first steps of Neil Armstrong during the Apollo 11 mission. The two astronauts continued to prepare the Rover, the vehicle which was pivotal to the success of the next three days of lunar exploration. While this was taking place, Fendell panned the camera to obtain a view of the landing area for the geologist team.

He explains, "The first thing they wanted us to do at each site was to do a thing called a pan…What we would do is move it [the TV camera] three degrees and this guy by the name of Tim (M.H.) Hait, who worked for the USGS, would take a Polaroid picture. And then we'd move it another three degrees and he'd take a picture. So by the time we did this pan, he had gotten a complete Polaroid view in that back room of that entire site that we were at, and they would lay that out and stick it all together, and we would start into following the crew and doing our things with the crew, because we knew where they were going, and they would start looking at what else they wanted to see."

Ed Fendell observing the Hadley base shortly before being handed control of the RCA TV camera. (NASA Photo S71-54920-34)

Ed Fendell at his INCO desk in Houston controlling the TV camera on the moon. (NASA Photo S71-54941-36)

Each station stop on the traverse made by the two astronauts in the Rover abided by Ed's site-documentation procedure, once they had set up the TV camera antenna. As the two astronauts explored the surrounding vicinity of their first stop, Elbow Crater, Ed Fendell used the camera to assist in the survey of the site. If a geologist needed to examine a rock or bolder up close, he would steer the camera and zoom in, allowing comprehensive documentation in the short minutes allotted to each stop. Although the other moonwalks had a TV camera allowing audiences to watch the activities live, on Apollo 15 the viewing experience became more like actually being on the lunar surface with the astronauts. The Hadley Mountains, the Rille and the surrounding area were all easily distinguished. When the camera panned to see its own shadow, the movement was quite noticeable as it moved either left or right. One problem, though, was becoming quite apparent, and that was that whenever the camera pointed towards the direction of the sun, the brightness washed out all other components of the scene. In a lot of these instances, the astronauts appeared extremely dark.

However, when there was no disruption from the sun, the views were stunning. At the next stop of Saint George Crater, even Capcom Joe Allen remarked, "We have a view of the Rille that is unearthly!" to which Jim Irwin replied, "Glad you can enjoy it with us."

Indeed, everybody was able to share in the wonder. Later, Dave Scott remarked, "Here sits this rock, and it's been here since before creatures roamed the sea on our little Earth." in reference to a rock he was about to examine. The two men continued to explore the region as fully as possible before Houston advised them to skip a further station stop and headed back to the Lunar Module to unload the ALSEP. Once they returned to the Lunar Module they began experiment package deployment, all the while being monitored by Houston and the plethora of people watching in their living rooms. They could also watch as core samples were taken from the lunar soil, and even observe, with concern, as the astronauts experienced problems removing the drill from it.

The whole time Houston closely monitored the activities often being able to supply better advice due to the fact of being able to watch the activities live. Once they finished the ALSEP work, they then began the closeout of the first day's EVA. As had been predicted, the new TV camera had functioned superbly.

The next day, television coverage began with the same view of the Rover wheel which had been transmitted the day before. Unfortunately, once again the sun washed out much of the shots of the astronauts whenever they stood between the sun and the TV camera. Drilling the lunar soil caused further problems, and despite receiving assistance from the people in Houston, who were able to survey the activities live, little progress was made--at least for the short term. The next stop would provide the biggest amount of geological excitement of the whole mission.

"Guess what we just found?" Dave Scott asked unable to contain his enthusiasm. "Guess what we just found? I think we found what we came for!" he reaffirmed. The rock they had found was an anorthosite, which is a sample of the lunar crust from when the moon was formed at least 4 billion years ago. It was given the name "Genesis Rock" and proved to be the geological trophy of the mission. They continued to explore the area wasting none of their limited time in collecting and categorizing numerous diverse lunar rocks.

The world watched live as the astronauts discovered what became known as "Genesis Rock".

The iconic photograph of Jim Irwin standing next to the Stars and Stripes remains one of the most reproduced of the Apollo Project. After the finishing activities around the flag, the astronauts completed their closeout and ingressed back into the Lunar Module.

Following that intense work, the team returned to the drilling site once more to attempt to resolve the ongoing problems obtaining core samples of the moon. The drill removal proved to be one of the more exhausting aspects of all the EVAs on the mission. Additionally, the astronauts scooped up the lunar surface before heading back to the Lunar Module.

Once back at the landing site, the camera began to experience its first real problem. It got stuck, unable to move due to the clutch mechanism not operating correctly. Luckily due to the redundant design feature of TV camera controls, a situation like this could be easily remedied simply by having an astronaut physically move the camera back into correct position. While the crew worked around the MESA, they also prepared for the ceremony of raising the United States flag. Normally this was done at the start of the mission, but on Apollo 15 it had been moved back.

The third day's EVA was shortened due to the necessity to launch at the predetermined time. Again, while the astronauts carried out the check listed activities, Ed Fendell took control of the TV camera and afforded TV views of the Hadley Plain. They returned again to the ALSEP site and attempted more drilling work. At the end of the stop the TV camera remained on during the next drive on the rover. Although the camera was pointing down, the moving lunar surface could be clearly seen during the brief moments when tracking stations were able to obtain signal lock from the TV signal.

The Terrace stop at Hadley Rille helped demonstrate how perspective recognition was severely hampered in the lunar environment. As the astronauts ventured towards the edge of the Rille, it appeared on the video feed that they could fall in any second.

 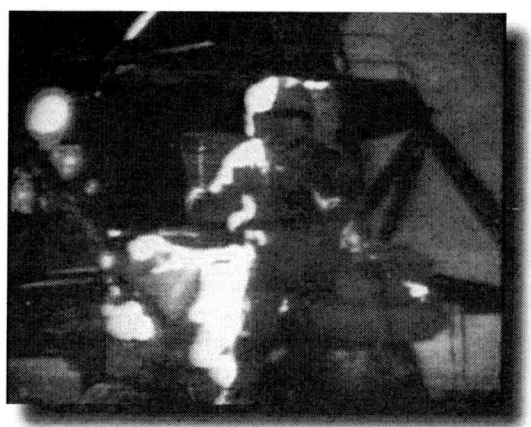

Ironically after returning closer to the Rover, Dave Scott lost his footing and tumbled over right in front of the camera. He quickly recovered and the work at the stops continued unabated. At the last stop prior to returning to the Lunar Module, the camera was stuck again, as it was looking up at the earth. It was returned to position but failed again once the astronauts had returned back to the landing site.

In the waning moments of the third EVA, Dave Scott set up an impromptu postal office on the moon, cancelling a first day cover on behalf of the U.S. Postal Service. Following that gesture he then demonstrated Galileo's principles of gravity in a vacuum by dropping a feather and a hammer at the same time. "How about that?" he proclaimed as they hit the surface at the same time, validating a centuries old theory.

The Rover was then driven to its final parking space several hundred feet away from the Lunar Module. The spacecraft sat on the screen unmoving as the voice communications of launch preparations could be heard.

The clutch system of the TV camera which malfunctioned near the end of the mission.

Unfortunately the clutch was not functioning properly and so it was unable to tilt the camera upwards. This was a crucial movement required to track the Lunar Module into the lunar sky as it flew into orbit. Nevertheless the camera was still operational and would afford viewers around the world with a spectacular lunar launch. As the launch time closed in, the camera view was unchanged, and suddenly a spray of Mylar and debris flew out as the Lunar Module ascended into the lunar sky. The disturbance to the surrounding area from the rocket exhaust was plainly visible on the TV feed. Due to the fuel mix being hydrazine, no visible flame was seen as the spacecraft flew upwards.

Following the astronauts' departure, the camera was still operational and continued to survey the landing site which was now missing the movement of the two astronauts. The view from the TV camera showed the remains of human visitation motionless on the Hadley Plain. While the TV was still active, several mountain features were also examined via the zoom lens although the lack of the tilt function did limit what could be shown. The opportunity of the solar eclipse was missed as the TV system on the Rover unexpectedly completely failed without warning. The television coverage from the moon's surface had come to a premature, yet overwhelmingly successful close.

TV transmissions from the Command Module continued on the long flight home. For the first time on an Apollo mission, the Command Module Pilot conducted a televised EVA to retrieve film from the mapping cameras on the outside of the spacecraft.

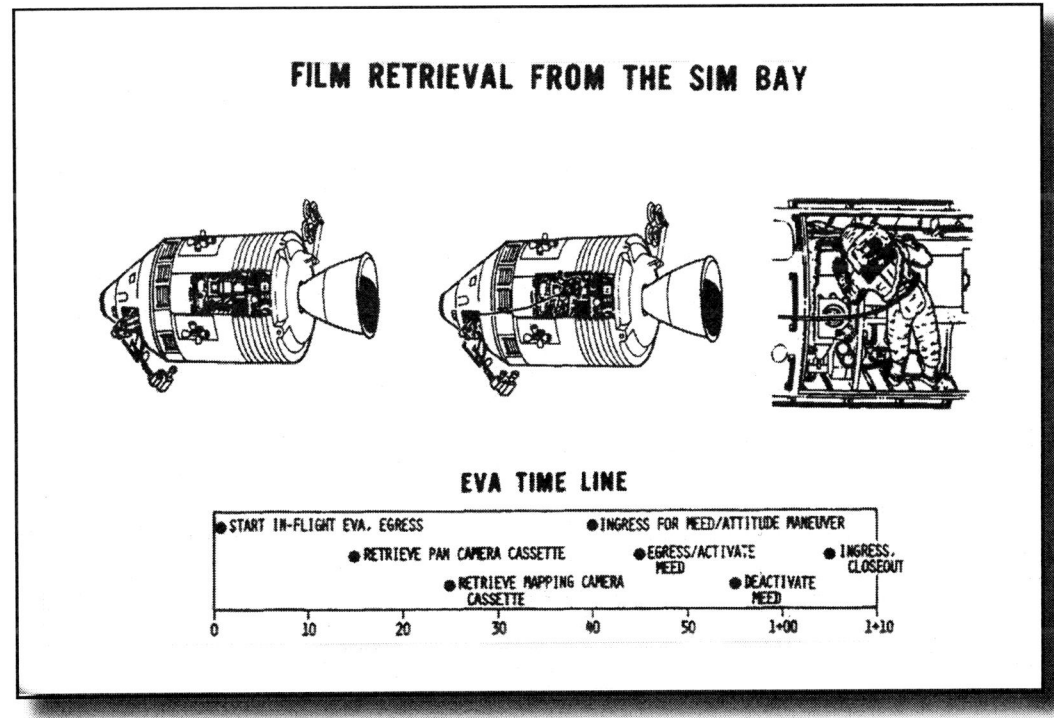

Press kit information on the film retrieval EVA.

During the spacewalk, the Lunar Module Pilot stayed at the hatch to oversee the procedure and set up both the film and TV cameras to record the event.

There was one telecast left to do before returning to earth. The final TV transmission was of the crew conducting the in-flight press conference. Three bearded astronauts sat at the lower equipment bay and answered questions given by the journalists gathered at Houston. They reflected on the success of the mission and wished the next crew all the success in the world.

The new television system had proved to the world that a compact system could work on the moon and be remote controlled from Houston. The standard had now been set.

COMPONENTS OF DELAY FOR REMOTE CONTROL OF TV CAMERA	Time (ms.)
MCC Processing of ECR	56
MCC-Goddard (GSFC) transfer	23
GSFC read-in at 50-kbps and processing time	25
GSFC - Goldstone transfer	35
or	
GSFC - Honeysuckle Creek,* Australia transfer via communications satellite	300
Remote Site Command Computer read-in at 4.8-kbps and processing time	152
Updata Buffer processing	30
Remote Site - Moon propagation delay (Average value, actually varies from 1200 to 1360)	1280
TV Camera Decoder read-in at 1.0-kbps	60
Moon - Remote Site propagation delay	1280
Goldstone - MCC transfer of TV picture	12
or	
Parkes* - MCC transfer of TV picture	300
Total using Goldstone	2953 or 3.0 sec.
Total using Honeysuckle Creek-Parkes	3506 or 3.5 sec.
Add a very small amount (at most 0.1 sec.) for TV picture processing en route.	

Calculated delay times for operation of the GCTA TV camera.

Television Event	GET HH:MM	GMT HH:MM	DATE GMT
1st Television Transmission START	3:25	16:34	26 July 1971
1st Television Transmission END	3:50	16:59	
2nd Television Transmission START	34:55:00	0:29	28 July 1971
2nd Television Transmission END	35:46:00	1:20	
3rd Television Transmission START	95:00:00	12:34	30 July 1971
3rd Television Transmission END	95:10:00	12:44	
EVA 1 Television Transmission START	119:52:00	13:26	31 July 1971
EVA 1 Television Transmission END	125:55:00	19:26	
EVA 2 Television Transmission START	142:35:00	12:09	1-Aug-71
EVA 2 Television Transmission END	149:20:00	18:54	
EVA 3 Television Transmission START	163:45:00	9:19	2-Aug-71
EVA 3 Television Transmission	168:20:00	13:54	
7th Television Transmission START	173:05:00	18:39	2-Aug-71
7th Television Transmission END	173:10:00	18:44	
8th Television Transmission START	173:35:00	19:09	2-Aug-71
8th Television Transmission END	173:40:00	19:14	
9th Television Transmission START	241:57:00	15:31	5-Aug-71
9th Television Transmission END	242:28:00	16:02	
10th Television Transmission START	270:22:31	20:56	6-Aug-71
10th Television Transmission END	271:12:22	22:08	

TV Transmission Schedule for Apollo 15.

CHAPTER 15. APOLLO 16

"Apollo 16 is gonna change your image!"

John Young on the Descartes region of the moon.

Apollo 16 was the second planned "J" mission which would again use the Lunar Rover, and the remote controlled TV camera. Original mission plans called for the first steps onto the lunar surface to again be televised from the stowage location on the MESA. Flight checklists reveal that the camera was then to be mounted on the tripod at a 12:00 position, again to televise the deployment of the Lunar Rover, as was the case on Apollo 15. In fact, the television component of the mission was very much a repeat of the procedures and coverage set on Apollo 15. The "being there" clarity of the GCTA pictures injected a newfound enthusiasm into general public opinion towards the missions.

Alan Bean caught wind of this rise in support and interest for the Apollo Project and composed a memo detailing his thoughts on how to better promote NASA on television. His idea was to have NASA spokespeople assist television commentators during quieter moments of the missions by highlighting how space exploration benefits the average American citizen. Bean was looking for any way possible for the astronauts themselves to help sell NASA and further missions, whether it be Skylab, or any new idea that NASA may have had for the future.

The immediate future however, was concerned with getting Apollo 16 to the Descartes region of the moon. There were a lot of unanswered scientific questions which, it was hoped, Apollo 16 would help to answer. The TV camera would once again aid the geologists in Houston as the mission was underway, to prepare exploration of interesting areas which could not have been anticipated until the Lunar module was on the moon. Bellcomm had further suggested modifications to the GCTA to allow it to observe low light phenomena such as the solar corona or zodiacal light, although it was well aware that any new proposals requiring camera modifications would more than likely not be ready in time for the remaining two missions.

However, some changes were implemented prior to the Apollo 16 flight. Although the TV camera had performed flawlessly on the previous mission, even when pointed in the direction of the bright sun, a precautionary item was included on Apollo 16 in the form of a lens hood.

(NASA Photo AS16-117-1876)

This also had the additional benefit of reducing glare on the lens (which then highlighted any accumulated lunar dust on the glass) anytime the sun came into shot. The clutch which had jammed on the last segments of the third EVA of Apollo 15 was completely redesigned to a much less complicated mechanism which proved via intensive testing to be extremely reliable even under significant temperature fluctuations. Indeed, the problem which hindered the televising of Apollo 15's Lunar Module launch did not present itself again on the two remaining lunar missions.

Several further changes to the GCTA were made to enhance expected performance on the moon. One modification was to reduce the "dead zone", which was the area to which the camera could not physically pan. The area which was a total of 20° on Apollo 15 was reduced to 12° on the remaining missions to allow the camera to view the astronauts while seated in the Lunar Rover. Other modifications pertained to thermal considerations for the TV camera on Apollo 16. The mirror surfaces on the TV system were improved to allow a reduction in heat build up during camera operation. The coating of the camera was also refurbished for the same thermal considerations.

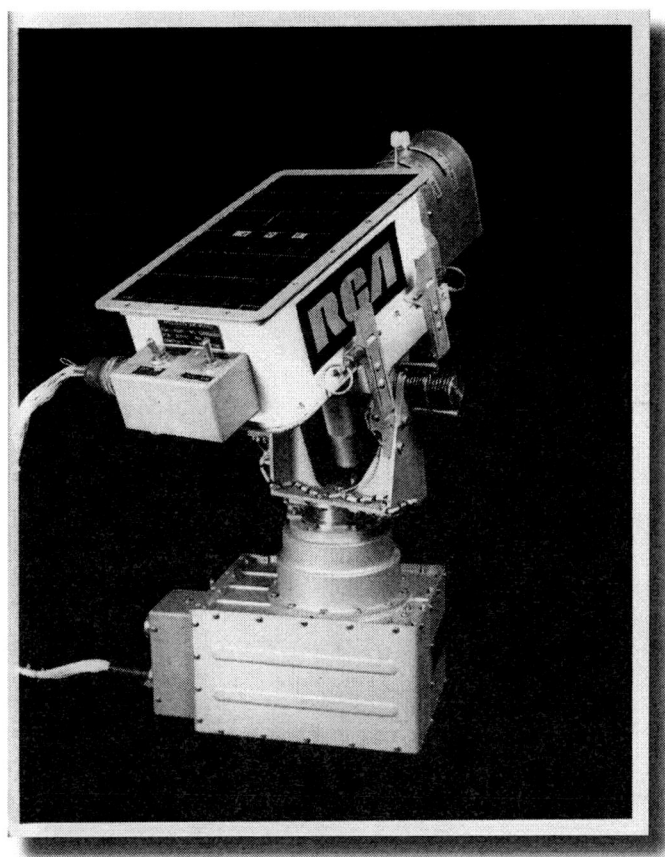

Promotional photographs of the RCA GCTA Camera. Photos courtesy Sam Russell.

For the Apollo 16 mission, the TV signal was to undergo image noise reduction before being converted to full color at Houston. John Lowry, who had just started his company, Image Transform, to transfer high quality copies of videotape over to film, had tested his algorithms on a segment of the Apollo 15 moonwalk. He was so impressed with the noise reduction on the signal that he contacted NASA to inform them of what was possible with his technology. Following a meeting with Jim McDivitt in January 1972, in which the demonstration material was played with a "before and after" comparison, Image Transform was contracted to enhance the TV signal for the two remaining lunar missions. The noise reduction worked by comparing static sections of a frame and removed the random noise patterns from the image. Sections of the frame showing motion were filtered, though not as thoroughly as the static sections. The end result was a cleaner image which also removed the noise introduced by the routing circuitry once the TV picture was received on earth, and it was done in near-real time! At NASA headquarters, a $46,008 contract was signed for Image Transform to process the incoming signal. The images from Apollo 16 onwards would be vastly superior to anything seen before.

Apollo 16 launched on April 16, 1972 with the crew of John Young as Commander, Ken Mattingly as Command Module Pilot, and Charlie Duke as Lunar Module Pilot. The fact that the lunar journeys were becoming more and more routine was partly responsible for the public's dwindling interest in the missions. Everything on the mission went smoothly until they were in lunar orbit. A problem with the Thrust Vector Controls on the Command Module "Casper" threatened to end the landing then and there.

The concern was that the problem would not allow the spacecraft to align properly during an engine burn. A fast paced work-around devised by engineers in Houston managed to side-step the issue, but due to the problem and the time required for the fix, the stay on the lunar surface was severely delayed. In an effort to reduce activities which would eat into the very precious EVA time the planned television coverage of the ladder descent of both astronauts from the MESA bound TV camera was dropped.

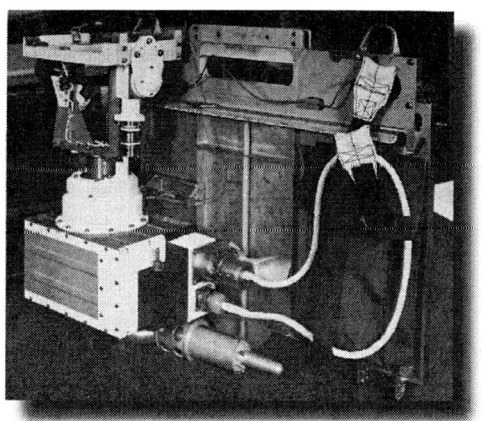

The TV camera control unit in its MESA stowage position.

So too was the placement of the camera on its tripod to monitor the deployment of the Lunar Rover. The television camera would begin transmitting only once it had mounted on the Lunar Rover. The first steps onto the moon were still to be recorded, albeit only with the 16mm Data Acquisition Camera fixed to the Lunar Module Pilot's window.

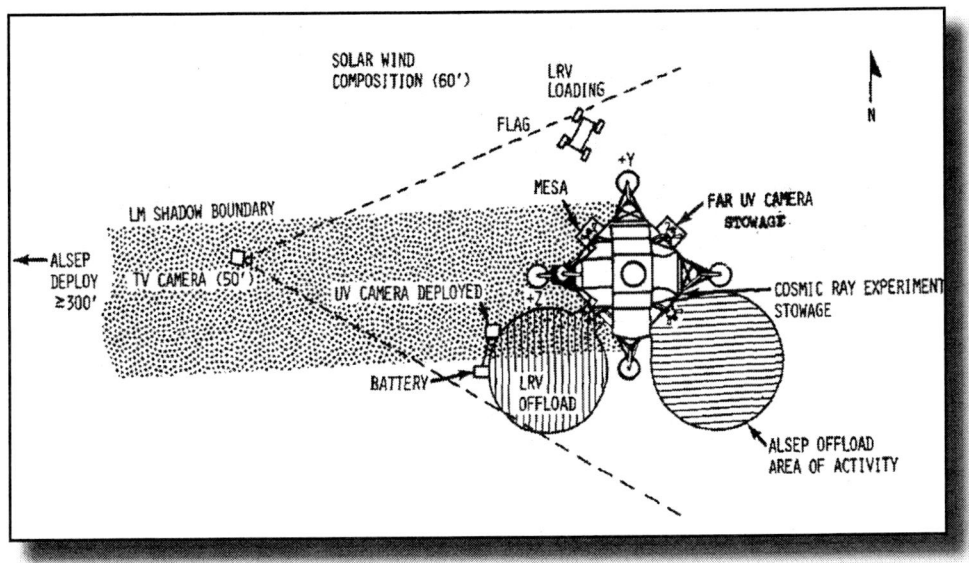

The original Apollo 16 lunar TV setup prior to mounting on the LRV which was ultimately discarded due to time considerations.

As had become standard procedure, Ed Fendell took control of the camera once it was mounted and activated on the Rover. He made a pan of the surrounding area of the landing site while the astronauts loaded up the vehicle. In one hilarious moment captured by the TV camera, Charlie Duke attempted to blow the dust off a film magazine cartridge. His embarrassment at catching himself is plainly evident in his body language while standing in front of the remote controlled camera. The new sunshade was then applied to the lens assembly and the results were immediate. No matter which direction the TV camera pointed, the image was hardly ever washed out as had been the case previously. John Young and Charlie Duke then assembled the American flag in front of the Rover's camera--John Young jumping as he saluted the flag!

During this procedure, Houston informed the two moonwalkers that Congress had approved funding for the Space Shuttle. At least for the time being, the future of NASA's space program seemed secure, despite the lunar missions soon coming to an end.

The two astronauts drove to the ALSEP site and immediately set up the experiments. While he was preparing a Heat Flow Experiment, Young accidentally tripped over the experiment's cabling and pulled the wiring out of its connection. He had effectively ruined the experiment live on television for the entire world to see and unfortunately Houston had not reacted fast enough to hinder the damage caused by the mission Commander. Transmission problems near Madrid caused a brief loss of picture for approximately four minutes on the incoming television signal. Undaunted, the two explorers continued to the scheduled stations for the first day's EVA.

The astronauts wasted no time at each stop in carrying out their geological survey. The TV camera reliably supplied Houston with panoramas of the surrounding area. The lens brush was used quite regularly to clean the moon dust off the TV camera lens during the traverse to each of the stops. There was a notable increase in picture quality each time the lens was cleaned, and the overall clarity of the picture was evident through the noise reduction implemented by Image Transform. As the two weary men closed out their day's exploration which had returned numerous and varied rocks, Fendell pointed the camera skyward and managed to show the earth hanging in the lunar sky. It was a profound moment where the world was looking back on itself from the surface of the moon.

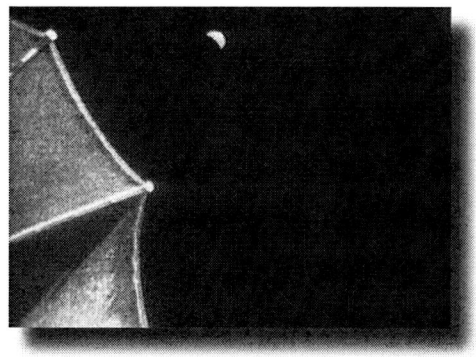

The next day, Young and Duke loaded up the rover and headed off again to continue their extensive examination of the geological surroundings. Unfortunately, after the magnificence of the Hadley Apennines, Descartes comparatively seems visually uninteresting as far as aesthetic beauty is concerned. As the Descartes region is predominantly highlands, there were few mountains to be seen, although North Ray Crater provided some unique close-up views of fissures on the moon. Duke himself commented that if he had fallen down the side of the crater, there was no way he would have been able to get back out! The TV system's ability to register finer details in the scenes it was shooting even prompted a remark by John Young of "Pretty good resolution, Tony!" as Houston advised him that he had just dropped some lunar samples without realising it. Thanks to the improved rotational field of view the camera could obtain, Charlie Duke was observed getting in to his seat on the Lunar Rover and strapping himself in as the astronauts prepared to drive to the next site.

Parking close to the Lunar Module, Young and Duke returned to the ALSEP site to continue their activities. This position afforded TV audiences some superb views of the LM in extremely good quality.

In fact they were close enough to the Lander that they could easily walk to it rather than drive. Although for ease of TV coverage and off–loading of equipment, the Rover was brought closer for the closeout of the EVA. Once again Fendell guided the TV camera to view the Earth in the Lunar Sky. There seemed to be a compulsion to repeatedly show the scene. It seemed only natural for humanity to want to steal the occasional peak at itself from such a distance.

The next day's EVA proved to be one that offered some of the most unusual television coverage of the mission. A field filled with rocks and boulders of varying size, in many cases the lack of atmospheric distortion made judging their sizes extremely difficult. One such boulder was what became known as "House Rock". Originally appearing in the distance on Fendell's preliminary pan of the area, it looked deceivingly small--perhaps half the size of the astronauts. As the men ventured closer to it, its relative size constantly changed. When they were approximately half way, it appeared to be twice as big as them. When the moonwalkers finally reached it, the boulder was huge, earning the name "House Rock".

The deceptive size of House Rock in comparison to the astronauts as they ventured closer to it.

After continuing their exploration of the lunar surface, the astronauts returned again to the Lunar Module to complete their visit on the moon. In a brief moment of spare time they displayed the effects of 1/6th gravity by jumping up and down. Charlie Duke attempted a jump, lost his balance, and landed flat on his back! For a moment he was concerned that he had damaged his suit, but all systems were still fully operational. As far as exciting TV was concerned that was the peak of activity for the lunar mission. John Young parked the Rover in its predetermined position to record the launch of the LM.

This time the clutch controlling the camera movement did not fail, and so an interesting view of the launch was envisioned. As the Lunar Module sat on the surface at Descartes, Fendell practiced the anticipated moves required to track the spacecraft upon launch. Operating purely on calculated times, he developed a process that would maximise the chances of tracking the Lunar Module for the maximum amount of time possible. The control commands would need to be made in Houston 1.5 seconds prior to the event taking place on the moon due to the time required for the signal to travel through space. If Fendell was using visual cues from the display monitor in Mission Control, he would have a delay of 3 seconds; therefore all his commands were timed according to the agreed launch time rather than from watching the incoming pictures of the launch. To compensate for the delay he also left a lot of space above the spacecraft in the frame so that if he was slightly late in tilting upwards the frame would still contain the Lunar Module on its ascent.

Upon launch this is exactly what happened, although as John Young had parked too close to the Lunar Module (80 meters instead of 100 meters) the camera could not tilt fast enough to keep the spacecraft in shot as it flew upwards. Upon pitch over it came back into shot but the difficulty in keeping it there was highlighted by the delay times. Several times the Lunar Module was found and subsequently lost again until finally, Fendell abandoned his continuing search and concentrated on the now abandoned lunar surface. Once again the eeriness of the familiar scene reverberated through Houston. Only the artefacts from the rush of activity which only hours earlier had been in full swing, were visible from the TV camera's perspective.

As far as the mission was concerned it wasn't over just yet. Several TV transmissions were now planned from the Command Module, the first being Ken Mattingly's half-hour EVA to retrieve the film canisters from the Command Service Module.

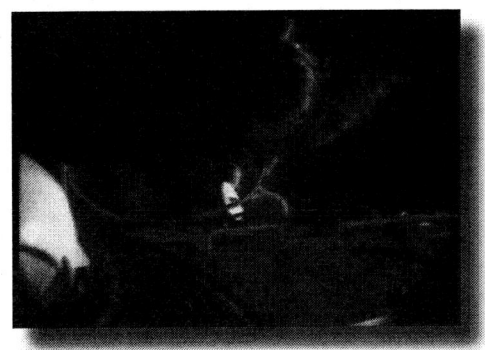

Further transmissions were made from inside the Command Module as the weary astronauts described and demonstrated aspects of their lives in space and gave their press conference.

For earth-based viewers it was a treat into a glamorous world of the men who were journeying to other worlds. Unlike the aviators or sailors of old, the TV camera allowed people to watch these voyages as they were happening. Unfortunately the interest of the people watching (or not watching to be more precise) was rapidly dwindling. Networks were not as committed to carrying uninterrupted coverage as in the groundbreaking days of the early Apollo flights. Such feats as staying on the moon for three days were being more frequently relegated to a highlights package shown on the evening news.

Television Event	GET Hh:mm	GMT Hh:mm	GMT Date
TV Transmissions 1 START	3:10	21:04	April 16 1972
TV Transmissions 1 END	3:28	21:22	
TV Transmissions 2 START	4:10	22:04	April 16 1972
TV Transmissions 2 END	4:20	22:14	
TV Transmissions 3 START	8:45	2:39	April 17 1972
TV Transmissions 3 END	9:06	3:00	
TV EVA 1 START	120:05:00	17:59	April 21 1972
TV EVA 1 END	125:35:00	23:29	
TV EVA 2 START	142:55:00	16:49	April 22 1972
TV EVA 2 END	149:40:00	23:34	
TV EVA 3 START	165:40:00	15:34	April 23 1972
TV EVA 3 END	171:10:00	21:04	
TV LM Launch START	175:15:00	1:09	April 24 1972
TV LM Launch END	175:40:00	1:34	
TV Transmission 8 START	176:18:00	2:12	April 24 1972
TV Transmission 8 END	176:25:00	2:19	
TV Transmission 9 START	202:57:00	4:51	April 25 1972
TV Transmission 9 END	203:12:00	5:06	
TV Transmission from LRV START	203:29:00	5:23	April 25 1972
TV Transmission from LRV END	204:12:00	6:06	
TV Transmission 11 START EVA	218:40:00	20:34	April 25 1972
TV Transmission 11 END	219:49:00	21:43	
TV Transmission12 START PRESS	243:35:00	21:29	April 26 1972
TV Transmission12 END	243:53:00	21:47	

TV Transmission times for Apollo 16.

CHAPTER 16. APOLLO 17

"Here men first completed the first explorations of the moon."

The plague attached to the leg of the Apollo 17 LM.

Apollo 17 was the 11th manned spaceflight of the Apollo program. It was also the last mission to journey to the moon, and the first to be launched at night. The mission launched at 12:33 a.m. on December 7, 1972 with the crew of Gene Cernan as Mission Commander, Ron Evans as Command Module Pilot, and Harrison "Jack" Schmitt as Lunar Module Pilot. Joe Engle's seat on the flight was bumped and given to Schmitt when it became clear the Apollo 17 mission was the last to go to the moon. The scientific community pressured NASA to send an astronaut scientist on the final mission and so Schmitt, the only qualified geologist with a hope of flying to the moon was taken from the cancelled flight of Apollo 18 and placed on Apollo 17 in place of Engle.

The Apollo 17 lunar landing site was the Taurus-Littrow valley. Based on lunar photography from earlier missions, this site was singled out as a locality where rocks both older and younger than those previously returned from other Apollo missions were expected to be found. Like the two missions which preceded it, the Apollo 17 mission was a "J" mission which featured the Lunar Roving Vehicle. Both the Westinghouse Command Module TV camera and the RCA GCTA TV system would be riding with the astronauts to document the last time humans would visit the moon in the 20th Century.

Due to the delayed launch, television coverage of the transposition and docking manoeuvre was not made. In fact no television was made on the journey out to the moon. The usual ingress into the Lunar Module and the guided tour that ultimately followed were notably absent from the television record of Apollo 17.

Following the routine voyage and orbit insertion around the moon, the Lunar Module "Challenger" detached from the Command Module "America" and flew to the landing site at the Taurus-Littrow Valley. The entire journey was a textbook case of space travel with nothing causing any major concerns. "Challenger" successfully landed and the two astronauts prepared to venture outside.

Like the previous flight of Apollo 16, the TV camera was not activated until the Lunar Rover had been successfully deployed. For this mission, however, it was planned that way prior to launch, as it did save valuable time from being wasted. "Hey it moves! It's alive!" said the astronauts as the remote control, camera came to life. Ed Fendell once again began the now familiar pan of the landing site to give the observers on the ground a basic idea of the vicinity around the spacecraft.

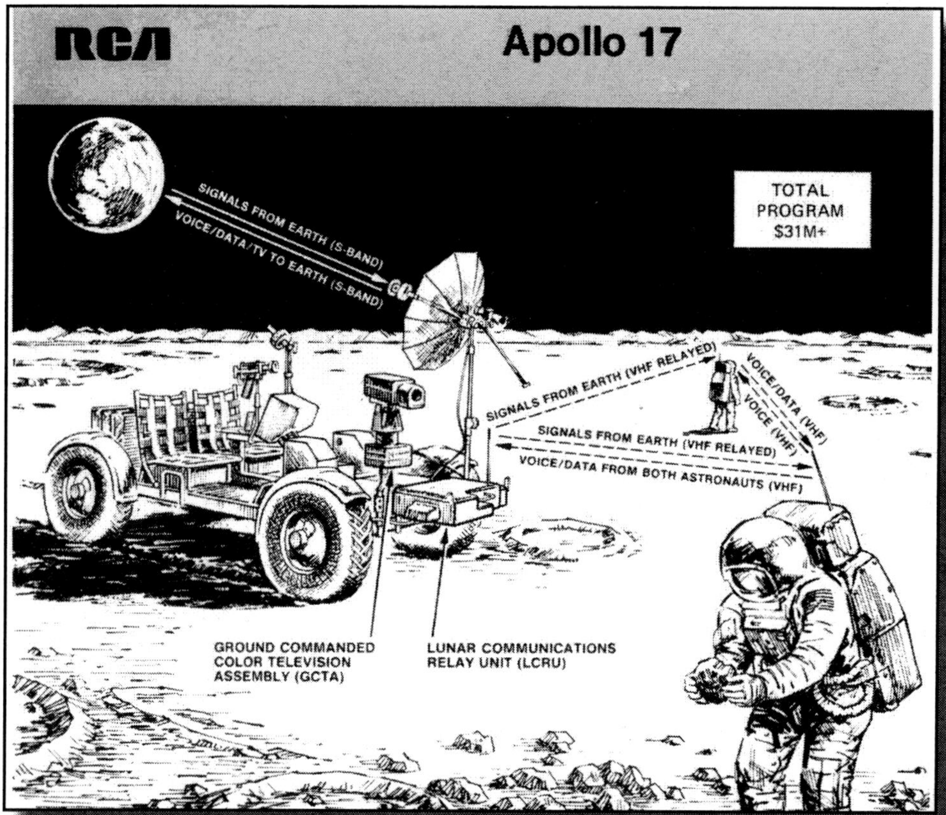

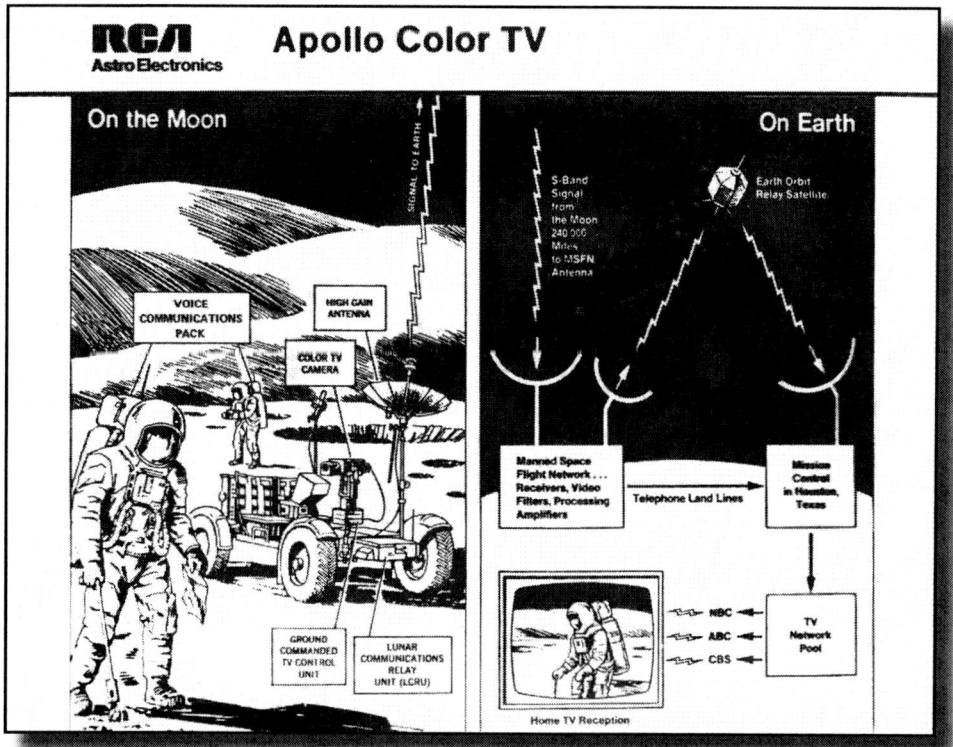

An illustration highlighting the color TV transmission from the lunar surface.
(RCA Astro Electronics Division Photos 72-9-520C and 73-3-508C)

The sun-proof tube proved itself as the camera panned past the sun. The iris closed down and the sun eventfully passed by without any effect on the camera other than some highlighting of the lunar dust on the lens. Cernan and Schmitt put up the American flag, which had been previously hanging in the Mission Control Room since Apollo 11.

It symbolically graced the scenery at the Taurus-Littrow landing site as the two astronauts posed for photographs of themselves, the flag and the planet Earth in the lunar sky.

Jack Schmitt carried items to the ALSEP site on foot highlighting the manner in which the lack of atmospheric hazing made judging of distances rather difficult. The further away he walked, the larger the surrounding details seemed to get, thanks to him acting as a reference point for scale on the valley floor.

Knowing full well who was controlling the camera, Cernan apologetically said, "Sorry about that Ed," as he swung the camera around and down into its stowed position prior to driving the Rover.

Television returned once again at the ALSEP site after Cernan realigned the antenna on the Rover. Both he and Schmitt set up all the experiments at the site including a Heat-Flow Experiment, study of the luna ejecta, seismic profiling, the lunar atmospheric composition and a gravimeter experiment. In the process of conducting the experiments, Jack Schmitt made a spectacular fall onto the ground, sending equipment sailing out slowly in the moon's 1/6th gravity. Further exploration of the site continued until time limits forced the two astronauts to move on to a geology site near Steno Crater.

The television panorama of the site revealed numerous craters visible on the side of the mountains flanking the valley region. Schmitt, with his geological training could hardly contain his enthusiasm as he scurried through the numerous rock specimens. The documenting and filing of the rocks sometimes made for not so exciting television, but for the other geologists observing the live feed, it was an amazing supply of data for them to make on-the-fly decisions about what to collect and whether extensions of stay were warranted.

The stop at the SEP site was made to set up an experiment to study the surface electrical properties of the moon. The site itself was only 70 meters from the Lunar Module, which could be seen standing isolated in the backdrop of the region's mountains. When the investigations were completed the crew ventured back to the Lunar Module parking the Rover nearby. An appraisal was made of the damaged fender on the Rover. Simulation teams, assisted by the television images would work out a temporary fix for the vehicle which the Apollo 17 astronauts were to carry out at the start of the next EVA. They unloaded the samples they had just spent the day gathering. Each time they approached the TV camera minor details on the suits highlighted just how good the television technology had progressed since the fuzzy black-and-white images from Earth Orbit aboard Apollo 7. If anything, the Apollo 17 mission only helped to reinforce just how much of a lost opportunity not sending another mission to the moon would be. Everything regarding lunar exploration was just starting to become efficient and the TV camera would have only gotten better over time.

After a well deserved rest the two lunar explorers loaded up the Rover once more, fixed the fender using a system of redundant maps and clamps from the Lunar Module, and headed towards more survey sites. At the next site, the dust on the lens had built up so much that the picture had degraded significantly. Cernan approached the TV lens and brushed away the dust all the while having his visor up so as to make his face inside the helmet clearly visible.

It was during this stop that Fendell began to explore the lunar sky, finding the earth as he did so, invoking cheers on the communications line picked up in Mission Control. The view was absolutely brilliant. It had been seen on previous missions, but on this one, it seemed almost poetic, like a beacon reminding the two men where they came from.

Station 3 at Lara crater revealed some of the most frustrating aspects of lunar exploration. Schmitt found the rock sampling using the restricted tools available to him quite difficult. At one point he fell in front of the TV camera prompting Capcom to remark that the Houston Ballet was looking for volunteers, dubbing Schmitt "Twinkle Toes" as he struggled to get to his feet. Cernan held the core sampling tube to the TV camera so as to let the geologists know that, even in the soft soil, only a very small amount of the core has been lost during removal. Moments like this highlighted just how much of a benefit a live TV signal could be in the real-time analysis of the moon. It would be the next Stop, though, which would supply the biggest "wow" factor of the mission.

Station 4 at Shorty Crater did not appear to hold any great mysteries when the astronauts first arrived. They began their routine survey with Ed Fendell doing the usual panorama of the site as they did so. It was Jack Schmitt's voice filled with excitement which made everyone of the geologists on Earth sit up and take notice.

Schmitt: "Oh, hey! Wait a minute..."

Cernan: "What?"

Schmitt: "Where are the reflections? I've been fooled once. There is orange soil!!"

Cernan: "Well, don't move it until I see it."

Schmitt: "It's all over!! Orange!"

Cernan: "Don't move it until I see it."

Schmitt: "I stirred it up with my feet."

Cernan: "Hey, it is!! I can see it from here!"

Schmitt: "It's orange!"

Cernan: "Wait a minute, let me put my visor up. It's still orange!"

Schmitt: "Sure it is! Crazy!"

Cernan: "Orange!"

The discovery potentially was an important one. As Schmitt himself explains, "The reason that this was a little more exciting than it might have even been otherwise was that, in our thinking about the origins of Shorty, we really had two major possibilities, with a number of observations related to each one. Number One was what we really believed and what turned out to be true: that Shorty was an impact crater that had penetrated the light mantle and was exposing the dark mantle and/or subfloor basalt beneath. The other alternative, just because of the lack of resolution and information from the pre-mission photographs, was that it was a dark-haloed volcanic crater; and, if that were true, one of the things that we had on our list to look for were signs of alteration, possible fluid-induced alteration of materials around the crater. And normally, on Earth, that alteration is colored, fumarolic alteration from oxidation. And so, when we saw orange, that was the immediate thought that everybody had. 'My God, there's a volcanic emanation here that's altered the soil.' Well, it turned out that that wasn't true, of course. It was volcanic material, but it was volcanic glass that had been spewed out of some fire-fountain-like eruptions 3.5 billion years ago that somehow had been protected from mixing with anything else, even though it was now at the surface. It had almost certainly been covered almost immediately by a lava flow, so that it was protected from meteor disruption and stirring. And then, when Shorty formed, somehow the pyroclastic ended up in the rim and a few other places in nearly pure form. There are some issues having to do with its origins that we can get to in a moment, but that's what caused all the excitement. We had sort of halfway predicted that we would find it there, even though it turned out not to be what we predicted we would find. They had a camera running in the Backroom, and what it shows is that, when that call came over, everybody sort of jumped up in the air and there was just total loss of control in the back science room."

The next stop at Station 5 detailed some very unusual boulder fields which even prompted Schmitt to remark to the TV camera, "Here I am folks, in the middle of a boulder field, just minding my own business."

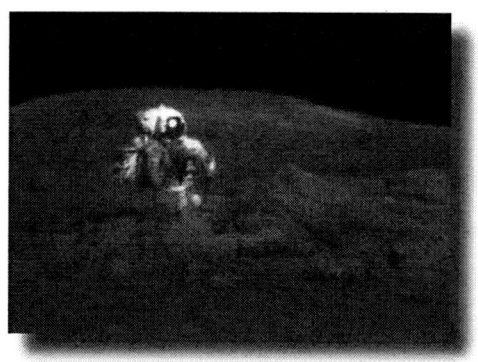

The exploration of the region gave the astronauts a large amount of samples, with the two men investigating the dark mantel of the sub-core around the crater. The stop was the last one before returning the Lunar Module for the day.

The crew unloaded the products of the day's exploration and Ed Fendell once more pointed the camera toward the Earth. For some reason the planet held a fascination with him which was not so evident on the other missions of Apollo 15 or 16. When he tilted the camera back down again to show the astronauts taking photographs of the area, the interactive aspect of the remote controlled camera came into play when Houston advised, "we can confirm that his lens cover's off." The attachment could be seen hanging off the end of the camera as Cernan made his set of photographs. Once the chores were completed, the two astronauts climbed back into the Lunar Module for a rest before their third and final EVA on the moon.

The next day's EVA progressed as normally and routinely as all before it. Fendell used the loading time to make panoramic sweeps of the Taurus-Littrow Valley, picking out interesting surface features to zoom in on and examine. Once again he captured stunning views of the Earth hanging above the astronauts. The two explorers then ventured on to Station 6 near the towering North Massif.

Orbital photography from the previous missions had revealed numerous boulders and boulder tracks which promised to be highly interesting to examine. The slope on which the Rover had parked made the two men appear to be standing at an unnatural slant. A huge boulder dwarfed the men as they moved up to it to collect samples from and around it. As Schmitt was climbing up part of the slope he turned and faced towards the TV camera. As his helmet visor had begun to scratch and thereby impair his sight, he opted to lift the protective gold layer. In one of the most iconic visual records of the entire Apollo program a human head peered through the impersonal spacesuit. It gave a face to the anonymous white figures walking in front of the camera. There were real men up there on the lunar surface.

Ed Fendell shot some of the valleys found near the site which appeared unearthly and mysterious, using the zoom lens to examine the intricacies of the view he was looking at via the TV camera. Ever drawn to the blue marble in the sky he pointed the camera for yet another Earth view as the astronauts loaded up the vehicle and headed off to two more stops before the end of their EVA.

There was more sampling of boulders and rocks. Attempts were made to roll one of them down the slope, but the age old rocks proved to be stubborn, prompting Schmitt to comment, "Go! Roll! Look I would roll on this hill, why don't you?"

They sampled the area underneath the moved boulder, and moved toward the Rover. Coming down the slope toward the end of the stop, Schmitt made skiing "whoosh" sounds and commented that he should have had his skis with him. At Station 9, a moment of curiosity struck Mission Control as the Explosive Package marker flag, from one of the deployed experiments passed the camera on one of Fendell's panoramas. For a brief moment not even having the TV camera zoomed in on it could assist them in figuring out what they were looking at. Luckily the astronauts explained what they were looking at and lifted the mystery of the object to the relief of all concerned.

Once back at the Lunar Module they unloaded the Lunar Rover and proceeded to stow all the items collected on their last venture on the Taurus-Littrow valley. They light-heartedly flung the redundant articles which were to be left on the moon, commenting on the distances the objects would fly.

Cernan: "Houston, before we close out our EVA, we understand that there are young people in Houston today who have been effectively touring our country, young people from countries all over the world, respectively, touring our country.

They had the opportunity to watch the launch of Apollo 17; (and) hopefully had an opportunity to meet some of our young people in our country. And we'd like to say first of all, welcome, and we hope you enjoyed your stay. Second of all, I think probably one of the most significant things we can think about when we think about Apollo is that it has opened for us - "for us" being the world - a challenge of the future. The door is now cracked, but the promise of the future lies in the young people, not just in America, but the young people all over the world learning to live and learning to work together. In order to remind all the people of the world in so many countries throughout the world that this is what we all are striving for in the future, Jack has picked up a very significant rock, typical of what we have here in the valley of Taurus-Littrow. It's a rock composed of many fragments, of many sizes, and many shapes, probably from all parts of the Moon, perhaps billions of years old. But fragments of all sizes and shapes - and even colors - that have grown together to become a cohesive rock, outlasting the nature of space, sort of living together in a very coherent, very peaceful manner. When we return this rock or some of the others like it to Houston, we'd like to share a piece of this rock with so many of the countries throughout the world. We hope that this will be a symbol of what our feelings are, what the feelings of the Apollo Program are, and a symbol of mankind: that we can live in peace and harmony in the future."

Schmitt: "A portion of a rock will be sent to a representative agency or museum in each of the countries represented by the young people in Houston today, and we hope that they - that rock and the students themselves - will carry with them our good wishes, not only for the new year coming up but also for themselves, their countries, and all mankind in the future."

Cernan: "We salute you, promise of the future."

They received final thanks from Dr James Fletcher, the Administrator of NASA. Cernan then got in the Rover moved it to its final parking position. The camera remained on during part of the drive allowing a brief look at the Rover in motion with the Lunar Module appearing for a moment in the background.

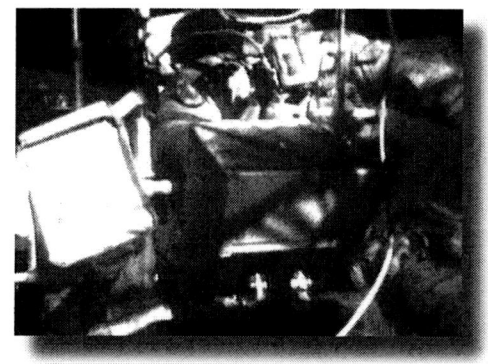

As he stopped the vehicle, Houston conveyed the concerns from the man running the TV camera.

Parker: "Okay. And Ed Fendell is hard on my back to remind you that it's better to be too far away than too close."

The further Cernan parked from the Lunar Module the easier it was for him to keep the spacecraft in frame when it launched. Having a wider shot would allow him more error in his timing should he not be completely accurate in starting his tilt up approximately 2 seconds before the ship actually launched. The reason being that there would be more of the frame the Lunar Module would be able to fly in before leaving the shot. On Apollo 16 the Rover had been left too close, resulting in only a few seconds of good footage after lift-off. On this final lunar launch, Fendell didn't want to take any chances. It had to work this time.

Once the two men were back inside the Lander, it was just a matter of adhering to the mission timeline before lift-off. Fendell did not watch the screen with the video image; he was glued to a stop watch which counted down to the launch. The shot he maintained prior to lift-off was a medium shot with the LM sitting at the bottom of the frame with a good deal of room above it. Approximately 3 seconds prior to ignition he pressed the "Zoom Out" key and the "Tilt Up" key.

As the LM moved upwards it was still in the camera's field of view. It was only once the LM was well on its way that Fendell dared zoom in, as the risk of losing the spacecraft was significantly reduced. As he had been controlling the camera for three days, he had familiarity with the delay and managed to keep the Lunar Module framed reasonably well for some time after launch.

The last view of the LM as it flew back to join the orbiting CSM.

When the spacecraft was no longer in sight, he tilted the camera back to the lunar surface which was now empty of life. All that remained from the human visitation were the artefacts left by the astronauts. As long as the antenna delivered a good signal to Earth, the TV camera was still used to good effect several more times after the LM launch.

As the two moonwalkers returned to the Command Module, the TV camera in that spacecraft transmitted their approach and docking to Earth. Just as they docked, Houston relayed a message from the President of the United States where he stated that the last exploration of the 20th Century had been concluded. Schmitt was livid. A whole generation of youngsters who dreamed of going to the moon had their dreams smashed, unnecessarily so.

There had been a plan to televise the impact of the Lunar Module onto the moon after it had been jettisoned from the Command Module. The Rover TV camera was positioned to the point which had been calculated by Mission Control as the impact site. Everything was in place, but unfortunately nothing was visible on the TV feed upon the moment of impact. What would have made absolutely spectacular live television was missed, but thankfully no part of the mission was compromised because of it.

Two more transmissions from the returning spacecraft were made. The first was the live telecast of Command Module Pilot Ron Evans retrieving the film canisters from the equipment bay outside of the Service Module. He waved to the television audience and conducted his spacewalk with glee.

The final transmission occurred with the regular press conference featuring all three astronauts answering questions asked by various journalists in Houston. There was a sense of sadness during the last TV signal from the spacecraft, despite the crew attempting to remain upbeat about the future of American space exploration. The biggest disappointment seemed to be that the United States was abandoning its lunar program just as it begun to get really, really good at it.

The in-flight press conference held by the last men to journey to the moon.

The flight of Apollo 17 set numerous records for the mission, the most rocks collected, the longest distance travelled, the longest stay on the lunar surface, and the longest mission to date. When the spacecraft landed in the Pacific Ocean the end of an era was final.

Suddenly it was all over. The footprints of Gene Cernan and Jack Schmitt have so far remained the last left on the moon by humans. When the audacious venture of sending men to the moon was envisioned over a decade earlier few could see the benefit or significance of taking a television camera on the flights. Even when the missions were underway, the television camera was seen as a hindrance, not as a benefit. What no-one at the time could have foreseen, simply because they were so busy working with their noses to the ground, was how much history would inevitably rely on the live transmissions to preserve the legacy of Apollo.

Television Event	GET Hh:mm	GMT Hh:mm	GMT Date
TV Transmissions 1 START	3:50	9:23	December 7 1972
TV Transmissions 1 END	4:10	9:43	
TV EVA 1 START	118:14:09	3:47:00	December 12 1972
TV EVA 1 END	123:32:09	10:05:00	
TV EVA 2 START	138:05:00	23:38	December 12 1972
TV EVA 2 END	144:55:00	6:28	
TV EVA 3 START	161:15:00	22:48	December 13 1972
TV EVA 3 END (including post launch TV)	unknown	unknown	
TV Transmission 4 START	234:10:00	23:43	December 16 1972
TV Transmission 4 END	234:35:00	0:08	
TV Transmission 5 START	254:54:00	20:27	December 17 1972
TV Transmission 5 END	255:42:00	21:15	
TV Transmission 6 START PRESS	281:20:00	22:53	December 18 1972
TV Transmission 6 END	281:47:00	23:20	

TV Transmission times for Apollo 17. Note: at the end of the third EVA the camera continued to send TV images long after the LM had left the lunar surface. Therefore, end of transmission times are unknown.

It is thanks to all these individuals that television was ready and included for the Apollo missions. Today, as the events slowly fade further and further away from the general population's collective memory, it is only through the TV records that any semblance of the excitement of the days of lunar exploration can be felt. Television allows an emotional connection between the viewer and the event, due in part to the association between TV and "live" that people harbour in their minds. The technology of videotape has an immediacy which allows events of long ago seem as if they were recorded only yesterday. The records of Apollo are more important today than they were as they were being made, for it is only through the television medium that new generations of space enthusiasts can be exposed to the fever and excitement of space exploration's golden era. The time when 12 people walked on the surface of the moon.

EPILOGUE

"...no one understood the extent of degradation due to conversion and transmission losses and as such they had no basis of comparison. They were just happy to have had any kind of video."
Stan Lebar

Back in 1969 as the first live slow-scan black-and-white images were arriving from the lunar surface, people were amazed at the technological feat allowing them to view humanity's first step onto another celestial body. In all the euphoria, it appears no-one stopped to think about either the historical importance, or the possibility of technological advancement which would occur in the future with regards to what they were watching.

The conversion process used to make the slow-scan low-resolution video compatible for regular TV sets used optical techniques mixed with a fledgling frame buffer system described throughout this book. The resultant signal introduced a significant amount of problems into the already weak signal. As a result, the footage the world recognizes as Neil Armstrong descending the ladder onto the lunar surface is one that is highly contrasted and extremely noisy.

At the time however, nobody gave the grainy images a second thought. America had achieved the impossible – landing humans on the moon and returning them to earth. Along the way a TV camera transmitted live pictures of the event and as far as everyone was concerned it had performed its duty.

It wasn't until 1997 that John Sarkissian, operations scientist at the Parkes radio Antenna in Australia, was queried by a British author as to the stations' role in Apollo 11. While researching which tracking station had indeed received the "first step" TV signal, he uncovered a Polaroid taken from the slow scan monitor as the live feed was coming in. The quality of the picture was remarkably better than any of the TV footage held by the NASA archives. Over time the desire to locate the original telemetry tapes increased. Several more photographs, as well as Ed von Renouard's Super-8 film of the slow scan monitors, emerged, revealing a far better quality of image than prior to the scan-conversion. From there the search for the original telemetry tapes began. The search team consisted of Stan Lebar, Dick Nafzger, John Sarkissian, Colin Mackellar and Bill Wood.

The telemetry tapes contained recordings of all the telemetry data from the LM, which included the information of the TV signal in its original raw form. All that was necessary was a tape machine capable of playing and decoding the signal and state-of the-art digital scan conversion which would eliminate all the extraneous noise introduced into the originally scanned video.

Despite a nearly six year world-wide search, the original tapes which could yield the better quality video signal have not been located. In a press conference held July 16, 2009 it was revealed that the tapes have been presumably wiped and re-used for later space mission data recordings. An experimental 2" tape recording system was used at Parkes during the Apollo 11 EVA to record the raw slow-scan television signal, but those tapes have also not been located.

However, the tape search did reveal a far better source of the EVA and of the all-important first step than what existed in the official archive. Lowry Digital restored the scan-converted video to a quality approaching that which would have been evident on the SSTV source. As of 2010, it is the best known source for the Apollo 11 EVA. Reflecting on future generations' appraisal of the first moonwalk, Stan Lebar lamented, "Those that follow us deserve better than what we had."

It is only in the last decade that a concerted effort has been mounted to present the video record of the moon landings in the best quality possible. As, sadly, those directly involved in the Apollo Project inevitably pass on, it is through the video record that future generations will be able to enjoy the lunar missions of NASA's golden age as those before them did.

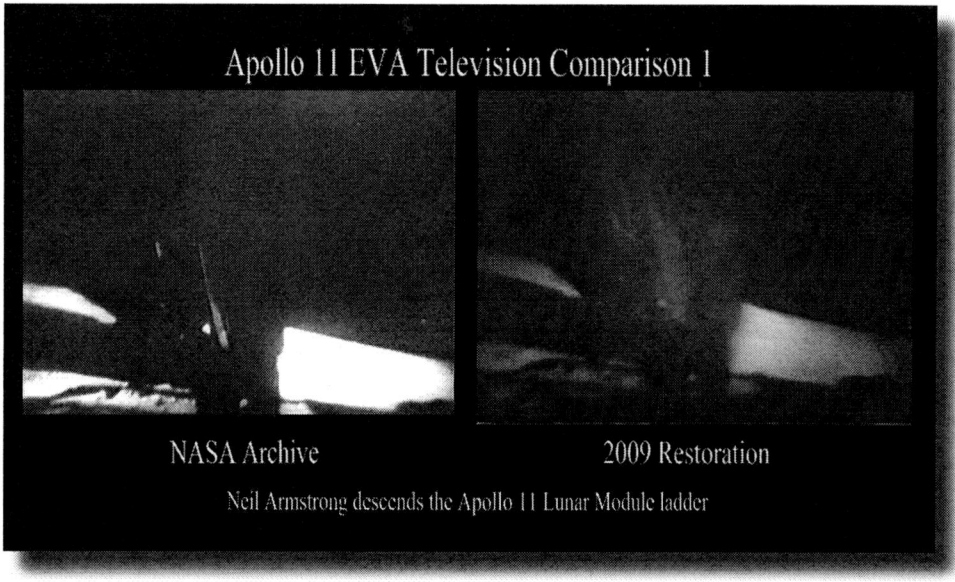

Apollo 11 EVA Television Comparison 1

NASA Archive — 2009 Restoration

Neil Armstrong descends the Apollo 11 Lunar Module ladder

BIBLIOGRAPHY AND REFERENCES

Abbey, George W. S.
"NASA response to citizen letter on TV failure on Apollo 12 letter"
December 8 1969
Record #40364, Program Apollo, Location 072-14, Johnson Space Center History Collection at University of Houston-Clear Lake.

Apollo 9 Mission Transcripts Apollo 9 Onboard Voice Transcription-Command Module, Vol. 1, April 1969.
[http://www.jsc.nasa.gov/history/mission_trans/AS09_CM1.PDF]

Apollo 9 Onboard Voice
Transcription-Command Module, Vol. 2, April 1969.
[http://www.jsc.nasa.gov/history/mission_trans/AS09_CM2.PDF]
Apollo 9 Onboard Voice Transcription-Lunar Module, March 1969.
[http://www.jsc.nasa.gov/history/mission_trans/AS09_LM.PDF]

Apollo 9 PAO Mission Commentary Transcript. March, 1969.
[http://www.jsc.nasa.gov/history/mission_trans/AS09_PAO.PDF]

Apollo 9 Technical Air-to-Ground Voice Transcript. March 1969.
[http://www.jsc.nasa.gov/history/mission_trans/AS09_TEC.PDF]

Apollo 9 Pre Flight Press Conference
From the Spacecraftfilms DVD set "Apollo 9: Spider Takes Flight"

Apollo 11 Technical Crew Debriefing
JULY 31, 1969
Prepared by: Mission Operations Branch
Flight Crew Support Division Vol. II

ARDEL
"Apollo TV Systems Reports 1965-1966 Lens seals"
1966
Report number 05035
Record #208687, Program Apollo, Location 083-61, Johnson Space Center History Collection at University of Houston-Clear Lake.

Bean, Alan
"Suggestion For our TV Support During Apollo 16" memorandum
August 17 1971

Bellcom
"Apollo 10 Color TV"
Report number B69-04030
April 7 1969

Bellcom
"Use of Goldstone Mars Station for Reception of LM Television from Lunar Surface"
Report number N79-72638
June 12 1968

Billings, Richard N., Schirra, Wally
"Schirra's Space"
Naval Institute Press, 1995
ISBN 1557507929, 9781557507921

Bissett-Berman Corporation
"Optimum Modulation and Multiplexing Techniques for Apollo Spacecraft to Ground Communications"
Report number Z65-10186
May 1 1962

Boeing Company
"Lunar Roving Vehicle Operations Handbook"
Report number NASB 25145
April 19 1972

Bohlmann, R.
GTCA Notebook
NASA KSC LS-ENG-52
April 13 1972

Borman, Frank in "Telecasts from Apollo 8" from the Race to the Moon information website of PBS.
[http://www.pbs.org/wgbh/amex/moon/peopleevents/e_telecasts.html]

Bottomley, T. A. & Schachne, S. H.
Bellcom
"LEM Critical Design Review"
Report number N78-75667
February 25 1966

Brooks, Courtney G., Grimwood, James M. and Swenson, Jr., Loyd S.
"Chariots for Apollo: A History of Manned Lunar Spacecraft"
The NASA History Series
Report number NASA SP-4205
Scientific and Technical Information Office
Washington, DC 1979

Calio, Anthony J.
"Television Requirements for Bldg. 17 Real Time Mission Support" memorandum
December 2 1971

Report number TF5/445-71
Record #44460, Program Apollo, Location 073-56,
Johnson Space Center History Collection at University of
Houston-Clear Lake.

Carrol, Parker
"Color Television System Costs" memorandum
May 14 1969
Record #37833, Program Apollo, Location 071-35,
Johnson Space Center History Collection at University of
Houston-Clear Lake.

Cernan, Gene interviewed during the CBS TV Coverage
of Apollo 10 May 1969
BBC Archives
Columbia Broadcasting System, Inc.
"10:56:20 PM, EDT, 7/20/69"
Columbia Broadcasting System, Inc.
1970

Corbett, Wayne W.
"Colored TV Presentation by RCA on June 11"
memorandum
June 12 1969

Cronkite, Walter
"The Flight of Apollo 8" CBS telecast
December, 1968
BBC Archives

Cronkite, Walter
Apollo 9 CBS launch telecast
March 3, 1969
BBC Archives

Cronkite, Walter
"The Flight of Apollo 10" CBS telecast
May, 1968
BBC Archives

Cronkite, Walter and Schirra, Walter
"Man on the Moon" CBS Apollo 11 telecast
July, 1969
BBC Archives

DATA System Division of Litton
"Scan Converter Feasibility"
Report number NAS 9 3512
February 8 1965

De Atkine, David and Merz, Paul H.
"Preliminary Alternate Mission and Abort Studies for
Apollo Missions AS-204A/5"
Report number 65-FM-126
September 15 1965
Record #25369, Program Apollo, Location 066-15,
Johnson Space Center History Collection at University of
Houston-Clear Lake.

Decker, J. L.
"Cancellation of Contract with NAA for Provision of a
CSM TV Camera for Apollo" memorandum July 9 1963

Record #19536, Program Apollo, Location 063-55,
Johnson Space Center History Collection at University of
Houston-Clear Lake.

Dietz, Reinhold H & Rhoades, Donald E.
"Apollo Experience Report - Lunar Module
Communications System"
Report number TN D-6974
September 1972

Di Prima C.D, Carlson H.
"Lunar TV Camera Optics Proposal"
Bell & Howell
Report number 7968
February 1965
Record #208688, Program Apollo, Location 083-61,
Johnson Space Center History Collection at University of
Houston-Clear Lake.

Drummond, John D.
"Color-wheel takes a spin in space."
Electronics Magazine July 7, 1969

El-Baz, Dr Farouq
Email correspondence regarding the LRV remote traverse
proposal.
April 1 and April 2, 2009

Engert, M.
"The Apollo Lunar Television System"
June 1 1965

Faget, Maxime A.
"LM Ascent TV Analysis" memorandum
March 20 1969
Record #37196, Program Apollo, Location 071-23,
Johnson Space Center History Collection at University of
Houston-Clear Lake.

Fairchild
"Detail Test Procedures for Apollo TV System"
Report numbers 008-0001-1 thru 008-0033-1
August 1 1965
Record #208683, Program Apollo, Location 083-62,
Johnson Space Center History Collection at University of
Houston-Clear Lake.

Fairchild
"Slow Scan TV Camera Monitor"
Report number NAS 9-1100
June 27 1966

Farmer, Norman B.
"Utilization of ALSEP Hardware in Support of a Lunar
TV System" memorandum
December 24 1969
Record #43032, Program Apollo, Location 073-16,
Johnson Space Center History Collection at University of
Houston-Clear Lake.

Fendell, Ed
"J Mission Lunar Surface TV Requirements Meeting

Minutes" memorandum
September 16 1970
Report number 70-FC24-248
Record #24278, Program Apollo, Location 072-63, Johnson Space Center History Collection at University of Houston-Clear Lake.

Flight Support Division MSC
"Network Controller's Mission Report Apollo 11"
August 15 1969

Garden, F. and Gilbert, A.
"Advanced Study of Video Signal Processing in Low Signal to Noise Environments Semi-annual Progress Report, 1967-1968"
November 1 1971

Gardiner, Robert A.
"Recommendations for TV Utilization During Apollo 9" memorandum
February 12 1969
Record #36724, Program Apollo, Location 071-12, Johnson Space Center History Collection at University of Houston-Clear Lake.

Gardiner, Robert A.
"LM Color TV Cameras" memorandum
September 29 1970
Record #42555, Program Apollo, Location 072-64, Johnson Space Center History Collection at University of Houston-Clear Lake.

General Mills
"Lunar TV Camera Manipulator Final Report"
Report number N63 21118
1963

GPC
"Minutes of Scientific Equipment Meeting No. 16 on TV Camera Design Interface"
April 22 1965
Record #24092, Program Apollo, Location 065-35, Johnson Space Center History Collection at University of Houston-Clear Lake.

Graier, James
"The Effect of Lunar Dust on EVA Systems During the Apollo Missions"
Report number TM 2005-213610
March 2005

Hall, D.
"Gyro-Dynamics Corporation Color System Proposal Problems" memorandum
April 13 1969
Record #37521, Program Apollo, Location 071-31, Johnson Space Center History Collection at University of Houston-Clear Lake.

Haney, Paul
"Apollo Television Capability on the Lunar Surface" memorandum
September 13 1967
Record #31005, Program Apollo, Location 068-53, Johnson Space Center History Collection at University of Houston-Clear Lake.

Haney, Paul. NASA Johnson Space Center Oral History Project
Oral History Transcript
Interviewed by Johnson, Sandra in High Rolls, New Mexico.
20 January 2003
[http://www11.jsc.nasa.gov/history/oral_histories/HaneyPP/haneypp.pdf]

Haronm, A. S.
"Effects of Parking Attitude on Thermal Performance of the LCRU and LRV Batteries"
Report number B72 02013
February 29 1972

Hayes International
"Apollo Logistics Support Systems MoLab Studies: TV Subsystem Studies for a Mobile Laboratory"
Report number N65-24018
March 31 1965

Hinners, N. W., El-Baz F. and Goetz A.F.H.
"A Preliminary ELM/Unmanned LRV Mission for the Appenine Front-Hadley Rille Area"
Bellcom report number N79-71819
May 31 1968

Heiberm, A.
"Feasibilty of High Gain Antenna Pointing System"
Report number B71 08021
August 17 1971

Hess, Wilmot N.
"Apollo Lunar Surface TV" memorandum
July 29 1968
Record #34342, Program Apollo, Location 075-26, Johnson Space Center History Collection at University of Houston-Clear Lake.

Hoffman, Phil
Private email exchange regarding the color switch for the sequential color switch.
August 4 and 5, 2009

Hogan, Alfred Robert
"Televising the Space Age"
2005

Hondros, George
"Considerations for Optimization of Ground System for Reception of Television from the Apollo Spacecraft at Lunar Distances"
Report number NASA TMX-5518
March 1 1965

Jayne jr, Allan W.
"Television and Video Glossary"
http://members.aol.com/ajaynejr

Johnson, J. E.
"Time Lag Associated with Remote Control of Apollo Lunar Surface TV Camera"
Bellcom report number B70-03083
March 30 1970

Johnston, Richard S.
"Design of a TV Camera by Westinghouse"
memorandum
July 14 1964
Record #21875, Program Apollo, Location 064-42, Johnson Space Center History Collection at University of Houston-Clear Lake.

Jones, Eric
TV Camera Lens Brush
All about the brush used to clean tv camera lens
23 April 2008

Kaminski, H
"Sternwarte Bochum beobachtet US-Apollo-Mondexperimente"
November 1972

Klaasen, K.P
"Utilization of LRV Power for the LCRU"
Bellcom report number B71 12004
December 1 1971

Kraft, Christopher
"Flight. My Life in Mission Control"
Plume February 26, 2002
ISBN-10: 0452283043 ISBN-13: 978-0452283046

Kraft, Christopher
"Onboard TV for AS-204" memorandum
May 9 1966
Record # 27152, Program Apollo, Location 066-65, Johnson Space Center History Collection at University of Houston-Clear Lake.

Kraus, H.
"Description of MSC/News Media Television Interface for Apollo 8 and 9"
Bellcom report number B69 05001
May 1 1969

Kuiper, Gerald P.
"Views Regarding TV Coverage of Future Apollo flights" letter
August 23 1969
Record #39213, Program Apollo, Location 071-56, Johnson Space Center History Collection at University of Houston-Clear Lake.

Lachenbruch, David
"Gotcha"
TV Guide
June 19 1971

Lachenbruch, David.
"The Raging Space-shoot Controversy".
TV Guide
May 10 1969

Lachenbruch, David
TV coverage of planned Apollo 16 mission.
TV Guide
April 15 1972

Lear Siegler
"Operation and Maintenance Manual: Slowscan Monitor Model 0762"
Report number 2-2676
January 27 1965

Lebar, Stan
"Lunar TV Camera: Statement of Work Exhibit A"
Westinghouse
August 15 1966

Lebar, Stan
emails (16, 17, 18 February 2009) discussing the development of the sequential color TV system for Apollo.

Lebar, Stan
private email discussing Westinghouse and the contract to supply the lunar TV camera.
25 February, 2009

Lebar, S. and Hoffman, C.
"TV Show of the Century: A Travelogue with No Atmosphere"
Electronics Magazine 1967
March 6 1967

Lee, William A.
"High resolution mode for Apollo TV camera"
memorandums
December 18 1964
Record #23004, Program Apollo, Location 075-13, Johnson Space Center History Collection at University of Houston-Clear Lake.

Lee, William A.
"The Case for Television Transmission During LEM Descent and Ascent" memorandum
May 27 1964
Record #21337, Program Apollo, Location 064-33, Johnson Space Center History Collection at University of Houston-Clear Lake.

Lindsay, Hamish
"Tracking Apollo to the Moon"
Praxis-Springer 2001
ISBN: 1852332123

Liwshitz, M. and Patterson, N. P.
"Use of LRV TV for Scientific Observations"
Bellcom report number B71 10011
October 19 1971

Lloyd, D.D.
"Apollo Lunar Surface TV System"
Bellcom report number TM 68-2011-2
April 15 1968

Lockheed Electronics
"Summary of Results Apollo 15 TV Configuration Optimization Tests"
MSC-EE7-274 (U)
July 7 1971

Lockheed Electronics Company
"Lunar Television Camera Pre-Installation Acceptance Test Plan"
Report number 28-105
March 12 1968

Lokke, B.H. and Cleveland, E.H.
"Apollo Spacecraft Familiarization"
Report number N74-72949
Spacecraft used in Apollo
December 1 1966

Low, George
"Television Camera to be flown in Apollo Spacecraft" memorandum
Report number PA-7-10-108
October 30 1967
Record #31454, Program Apollo, Location 068-64, Johnson Space Center History Collection at University of Houston-Clear Lake.

Low, George
"Lunar Surface TV Simulation Test Report"
Report number PA-9-6-68
June 26 1969
Record #38445, Program Apollo, Location 06071-45, Johnson Space Center History Collection at University of Houston-Clear Lake.

Low, George M.
"Color TV Costs"
May 10 1969
Record #37784, Program Apollo, Location 071-34, Johnson Space Center History Collection at University of Houston-Clear Lake.

Low, George M.
"Status of Apollo 10 Color TV System" memorandum
April 9 1969
Record #37393, Program Apollo, Location 071-25, Johnson Space Center History Collection at University of Houston-Clear Lake.

Lowrance, John
Email correspondence regarding the Block I RCA TV camera.
December, 2008.

Lowrance, J.L. and Zucchino, P.M.
"Television Camera System for the Command Module of the Apollo Spacecraft"
SMPTE Journal Vol. 74, February 1965

Lovell, James. NASA Johnson Space Center Oral History Project
Oral History Transcript
Interviewed by Stone, Ron in High Rolls, New Mexico.
25 May, 1999
[http://www.jsc.nasa.gov/history/oral_histories/LovellJA/lovellja.pdf]

Lunde, Alfred N.
"GCTVA View of the Solar Eclipse from Hadley Rille" memorandum
July 26 1971
Record #43997, Program Apollo, Location 073-43, Johnson Space Center History Collection at University of Houston-Clear Lake.

Manned Spacecraft Center
Apollo 8 Onboard Voice Transcription,
As Recorded on the Spacecraft Onboard Recorder (Data Storage Equipment)
January 1969

Martin Marietta Corporation
"Potential Applications of Digital Techniques to Apollo Unified S-Band Communication System"
Report number N70-42895
February 1970

Mason Dixon Astronomer
Newsletter detailing anecdotes of the Westinghouse/RCA war.
August 2005

Mason, Ed
"Apollo 12 TV"
October 3 1969
Record #39557, Program Apollo, Location 071-64, Johnson Space Center History Collection at University of Houston-Clear Lake.

McClanahan, J. T.
"Apollo TV System Briefing of MSC Current Planning" memorandum
June 10 1966
Record #27374, Program Apollo, Location 067-14, Johnson Space Center History Collection at University of Houston-Clear Lake.

McDivitt, Jim
"Remote Control Capability for Lunar Surface TV"
Report number PA/RWK-M106
December 19 1969
Record #40527, Program Apollo, Location 072-16, Johnson Space Center History Collection at University of Houston-Clear Lake.

McDonnell Aircraft Corporation
"Direct Flight Apollo Study."
Report number C-49069
October 3, 1962

McDonnell Aircraft Corporation
"Direct Flight Apollo Study. Vol. 1: Two Man Apollo Spacecraft."
Report number C-72078
October 31, 1962

Menard, J. Z.
Requirements for Lunar Surface TV for Apollo 16 and Subsequent Missions
Bellcom report number B69 11051
November 19, 1969

Mengel, J.T., Vonbrun, F.O. and Varson, W.P.
"An Integrated NASA Tracking Network for Apollo"
Report number TM X-55937
August 23, 1962

Messner, Max H.
"The Television Camera System Used in Apollo 7 and 8 Command Modules"
SMPTE vol. 79 no. 1
1970

Mindell, David A.
Digital Apollo. Human and Machine in Spaceflight "
MIT Press
ISBN-10: 0262134977
ISBN-13: 978-0262134972
2008

Mission Evaluation Team
"Lunar Roving Vehicle/Traverse Gravimeter Experiment Motion Sensitivity Test"
Report number N73-25882
April 1973

Morse jr, A. E.
"LM-6 color television test" memorandum
Report Number PSK-9M-254
November 5 1969
Record #480873, Program Apollo, Location 079-15B FF9,
Johnson Space Center History Collection at University of Houston-Clear Lake.

MSC
"Apollo AS-204A Flight Plan"
December 9 1966
Record #209640, Program Apollo, Location 077-24, Johnson Space Center History Collection at University of Houston-Clear Lake.

MSC
"Apollo 15 Temporary Loss of Command Module Television Picture"
Report number JSC-07912
March 1973

MSC
"Apollo 9 Mission Report Supplement 11 Communications System Performance"
Report number PA-R-69-2
December 1969
Record #40624, Program Apollo, Location 078-32, Johnson Space Center History Collection at University of Houston-Clear Lake.

MSC
"Apollo 16 Malfunction of Television Camera Monitor on Command Module"
Report number MSC-07489
October 1972
Record #45540, Program Apollo, Location 080-24, Johnson Space Center History Collection at University of Houston-Clear Lake.

MSC
"Final Photographic and TV Operations Plan Apollo 8"
November 18 1968
Record #209525, Program Apollo, Location 077-66, Johnson Space Center History Collection at University of Houston-Clear Lake.

MSC
"Final Photographic and TV Operations Plan Apollo 12"
October 20 1969
Record #209400, Program Apollo, Location 079-14, Johnson Space Center History Collection at University of Houston-Clear Lake.

MSC
"Preliminary Photographic and TV Procedures Apollo 13"
January 15 1970
Record #40840, Program Apollo, Location 079-42/43-A, Johnson Space Center History Collection at University of Houston-Clear Lake.

MSC
"Final Photographic and TV Procedures Apollo 13"
April 3 1970
Record #41442, Program Apollo, Location 079-42/43-A, Johnson Space Center History Collection at University of Houston-Clear Lake.

MSC
"Apollo 15 (J Mission) Design Certification Review"
March 17 1971
Record #209112, Program Apollo, Location 079-62, Johnson Space Center History Collection at University of Houston-Clear Lake.

MSC Final
"Apollo 11 Lunar Surface Operations Plans"
June 27 1969
Record #438483, Program Apollo, Location 078-46, Johnson Space Center History Collection at University of Houston-Clear Lake.

MSC NASA
Roundup: MSC Newsletter
vol 9. no 3.
November 28 1969

MSC NASA
Roundup: MSC Newsletter July 1969
vol. 8. no. 19
July 11 1969

Mudgway, Douglas J.
"Big Dish. Building America's Deep Space Connection to the Planets"
University Press of Florida
2005

Mueller, George E.
"Apollo Lunar Exploration Missions - Real Time TV Support Requirements" memorandum
November 6 1969
Record #40004, Program Apollo, Location 072-12
Johnson Space Center History Collection at University of Houston-Clear Lake.

NASA
"Apollo 7 Press Kit"
October 6 1968

NASA
"Apollo 8 Press Kit"
December 15 1968

NASA
Apollo 9 Press Kit
February 23 1969

NASA
Apollo 10 Press Kit
May 7 1969

NASA
"The Apollo Spacecraft - A Chronology Vol II"
December 8 2007

NASA
"Slow Scan Converter"
conglomerated reports
September 15 1968
Center Series, Communication Documents,
Section 109 Box Number: 04,
Johnson Space Center History Collection at University of Houston-Clear Lake.

NASA
"Westinghouse Scan Converter"
conglomerated reports
February 19 1969
Center Series, Communications Documents,
Section 109 Box Number: 04,
Johnson Space Center History Collection at University of Houston-Clear Lake.

NASA
"Lunar TV Development Contract"
conglomerated reports
July 3 1967
Center Series, Communications Documents,
Location: Section 109 Box Number: 04,
Johnson Space Center History Collection at University of Houston-Clear Lake.

NASA
"Apollo 12 Color TV Test Summary
Conglomerated Reports"
November 20, 1969
Center Series, Communications Documents,
Location: Section 109, Box Number: 11,
Johnson Space Center History Collection at University of Houston-Clear Lake.

NASA
"LCRU/LSS TV Preliminary Tests"
Conglomerated Reports
December 1969
Center Series, Communications Documents,
Location: Section 109, Box Number: 11,
Johnson Space Center History Collection at University of Houston-Clear Lake.

NASA
"LCRU Pre/Deemphasis Tests"
Conglomerated Reports
May 15, 1970
Center Series, Communications Documents,
Location: Section 109 Box Number: 11,
Johnson Space Center History Collection at University of Houston-Clear Lake.

NASA
"LCRU Engineering Model"
Conglomerated Reports
1970
Center Series, Communications Documents,
Location: Section 109, Box Number: 11,
Johnson Space Center History Collection at University of Houston-Clear Lake.

NASA
"LM Color TV Tests"
Congomerated reports
1969
Center Series, Communications Documents,
Location: Section 109, Box Number: 11,
Johnson Space Center History Collection at University of Houston-Clear Lake.
NASA
"Crew Training Manual: LM Communications Relay Unit" (Excerpts)
Report number LRTM-SY 1 REV D
March 30 1971

NASA
"Report on Apollo 13 Review Board: Appendix A: Baseline data Apollo 13 Flight System Operations"
Report number N70-78507
1970

NASA
"Apollo 13 Press Kit"
1970

NASA
"Apollo 14 Press"
1971

NASA
"Lunar TV Reliability Evaluation Plan"
NAS 9-3548
April 15 1965

NASA
"Final Report for Color TV Study"
Report number NAS 9-5342
Report on Color TV from Nov 1965 to Mar 1966

NASA News Releases from 1964 - 1975
Various news Reports featuring updates on TV camera development.

NASA
"Apollo Unified S-Band Technical Conferences"
Report number SP-87
July 14 1965

NASA
"Manned Space Flight Network"
Report number TMX 66479
November 1970

NASA
"Apollo Experience Reports: Lunar Module Communications System"
Report number TN D6974
September 1972

NASA
"Apollo Experience Report"
Report number TN-D7476
November 1973

NASA Electrical Systems Board.
"Thoughts on a TV System."
Originator not specified - or unknown and not determinable from document.
January 23, 1962
Record #17105, Program Apollo, Location 062-56,

Johnson Space Center History Collection at University of Houston-Clear Lake.

Niemyer, jr., L.L.
"Apollo Color Television Camera"
September 16 1969

Niemyer, L.L., Lebar, S.
"Testing of Apollo Black and White Television Cameras"
Report number NAS 9-10997
August 12 1971

Niemeyer, L.L. and Svensson, E.L.
"Apollo Television Cameras"
SMPTE Journal
vol. 79 Oct 1970

North American Aviation
"Apollo Operations Handbook"
Report number N73 70287
1966

North American Aviation, Inc.
Apollo Monthly Progress
Report number SID 62-300-1
31 January 1962

North American Aviation, Inc.
Apollo Monthly Progress
Report number SID 62-300-2
28 February 1962

North American Aviation, Inc
Apollo Monthly Progress
Report number SID 62-300-3
30 April 1962

North American Aviation, Inc
Apollo Monthly Progress
Report number SID 62-300-4
31 May 1962

North American Aviation, Inc
Apollo Monthly Progress
Report number SID 62-300 6
31 August 1962

North American Aviation, Inc
Apollo Monthly Progress
Report number SID 62-300-11
1 April 1963

O'Conner, J. J.
"Moon-to-Earth Antenna Look Angles for Continuous TV Coverage During LRV Traverses"
Bellcom report number B71 08021
August 18 1971

O'Conner, J. J.
"On the Problem of Continuous Television During Rover Traverse"
Bellcom report number B71 08036
August 26 1972

Ohnesorge, Thomas E.
"Apollo Experience Report: Electronic Systems Test Program Accomplishments and Results"
Report number NASA TN D-6720
March 1972

Orloff, Richard W. and Harland, David M.
Apollo: The Definitive Sourcebook
Springer Praxis Books
April 27, 2006
ISBN-10: 0387300430, ISBN-13: 978-0387300436

Painter, John H. and Hondros, George
"Unified S-Band Telecommunications Techniques for Apollo"
Report number TN D-2208
March 1965

Peltzer, K. E.
"Apollo Unified S-Band System"
Report number TM X-55492
April 1966

Petrone, Rocco A.
"Definition of Requirements for and Implementation of TV in LM10 and Subsequent Missions" memorandum
October 7 1969
Record #39592, Program Apollo, Location 075-36, Johnson Space Center History Collection at University of Houston-Clear Lake.

Petrone, Rocco A.
"LCRU and TV Development" memorandum
December 23 1969
Record #40576, Program Apollo, Location 072-16, Johnson Space Center History Collection at University of Houston-Clear Lake.

Petrone, Rocco A.
"Real Time TV Support During Apollo 12" memorandum
October 14 1969
Record #39664, Program Apollo, Location 071-65, Johnson Space Center History Collection at University of Houston-Clear Lake.

Phillips Samuel C.
"Apollo On Board Color TV" memorandum
June 13 1969
Record #38233, Program Apollo, Location 071-43, Johnson Space Center History Collection at University of Houston-Clear Lake.

Phillips, Samuel C.
"Apollo On-Board TV" memorandum
April 10 1968
Record #33115, Program Apollo, Location 069-53, Johnson Space Center History Collection at University of Houston-Clear Lake.

Phillips, Samuel C.
"Apollo On Board TV" memorandum
December 16 1968
Record #36048, Program Apollo, Location 070-54, Johnson Space Center History Collection at University of Houston-Clear Lake.

Phillips, Samuel C.
"Apollo 8 TV Implementation Plan"
December 2 1968
Record #35898, Program Apollo, Location 070-53, Johnson Space Center History Collection at University of Houston-Clear Lake.

Prokop, E.J.
"Apollo Slow Scan TV Signal Enhancement"
January 1 1970

Putman, R. E.
"The Role of Film and Television Technology in Space Achievements"
SMPTE vol. 79 no. 1
January 1970

Quinn, M.J. jr.
Apollo Slow Scan TV Transmission Tests Over Commercial Long
February 2 1967

RCA
"RCA Operations and Maintenance Manual: Apollo Command Module (CM) Television Camera Subsystem"
Report number AED M-2182
April 7 1969
Center Series, Communications Documents,
Location: Section 109, Box Number: 04
Johnson Space Center History Collection at University of Houston-Clear Lake.

RCA
"RCA Familiarization Manual: Apollo Command Module (CM) Television Camera Subsystem"
Report number AED M-2183
May 5 1969
Center Series, Communications Documents,
Location: Section 109, Box Number: 04
Johnson Space Center History Collection at University of Houston-Clear Lake.

RCA
"Operational Manual for the Apollo Command Module (CM) Television Camera"
Report number M-2155
July 29 1968
Center Series, Communications Documents,
Location: Section 109, Box Number: 04
Johnson Space Center History Collection at University of Houston-Clear Lake.

RCA
"Ground Commanded Television Assembly: Interim Report"
Report number N73-19192
February 15 1972

RCA
"RCA Ground Commanded Television Assembly (GCTA) Operation and Checkout Manual"
Report number NAS 9-11260
May 24 1971

RCA
"Apollo Command Module TV Camera"
Report number PO-AF-1079
March 1 1968
Center Series, Communications Documents,

Location: Section 109 Box Number: 04
Johnson Space Center History Collection at University of Houston-Clear Lake.

RCA
"Ground Commanded Television Assembly"
Report number R3901 F
December 29 1972

Rector, W. F.
"Lunar TV Cable" memorandum
April 22 1965
Record #24087, Program Apollo, Location 065-35,
Johnson Space Center History Collection at University of Houston-Clear Lake.

Rector, W. F.
"Addition of TV Cameras to Exhibit D, GFP/CFE List"
January 11 1965
Record #23167, Program Apollo, Location 065-12,

Johnson Space Center History Collection at University of Houston-Clear Lake.

Rosenblum, I. I.
"Comparison of Measured LM/ EVA Link Transmission Losses on Apollo 15 With Predicted Values"
Bellcom report number B71-12012
December 20 1971

Rusnak, Kevin M.
"Ed Fendell Interview transcript"
NASA Oral History Project
October 19 2000

Russell, Sam
Email correspondence regarding the RCA GCTA TV camera, between 2005 and 2009.

Russell, Sam
"RCA Camera Development"
in Space For Mankind's Benefit
Report number N73-13829 through N73-13874
November 19 1971

Russell, Sam
"Shooting the Apollo Moonwalks"
Internet based Essay by Sam Russell regarding the RCA camera
www.russelland.com

Sagan, Carl
referring to the photograph known as "The Pale Blue Dot" and excerpted from a commencement address delivered May 11, 1996

Sarkissian, John
"On Eagle's Wings: The Parkes Observatory's Support of the Apollo 11 Mission"
vol. 18 2001

Sawyer, Ralph S.
"Lunar Television Camera Weight Revision"
June 12 1965
Record #24282, Program Apollo, Location 075-14,

Johnson Space Center History Collection at University of Houston-Clear Lake.
Schilling, Donald L., Clarke, Kenneth K. and Pickholtz, Raymond L.
"A Space Communications Study"
Report number N 6926355
March 15 1969

Schirra, Walter M. Oral History Transcript
Interviewed by Neal, Roy in San Diego, California.
1 December 1998
[http://www11.jsc.nasa.gov/history/oral_histories/SchirraWM/schirrawm.pdf]

Schmid, P. E.
"The Feasibility of a Direct Relay of Apollo Spacecraft via a Communication Satellite"
Report number TN D 4048
August 1967

Schmid, P. E.
"Lunar Far-Side Communication Satellites"
Report number NASA TN D-4509
June 1968

Schraber, Gerad
"The US Geological Survey, Branch of Astrogeology - A Chronology of Activities from Conception Through the End of the Project"
2005-1190 Open File Report
2005

Schroeder, N. W.
Effects of Deletion of LM Erectable Antenna on LM-MSFN Communications
Bellcom report number B70 09082
September 30 1970

Schroeder, N. W.
"Performance of Communications Link Between Apollo CSM and DSN 210ft Antenna Goldstone"
Bellcom report number B69-05033
May 9 1969

Seamans, Robert C.
"Potential TV Coverage on Apollo" memorandum
plus follow-up related memorandums
March 30 1966
Record #26862, Program Apollo, Location 066-61,

Johnson Space Center History Collection at University of Houston-Clear Lake.
Selden, R.L.
Anticipated Performance of the Apollo 10 Color Television System
Bellcom report number N79-72666
April 21 1969

Selden, R. L.
"Alternative Design for the Television Transmission System of the Lunar Communications Relay Unit (LCRU) and their Communication Link Performance"
Bellcom report number B70 02039
February 4 1970

Seldon, R. L.
"Communications from the Lunar Surface to the MSFN without the Use of the LEM Erectable High Gain Antenna"
Report number X66 36747
April 27 1966

Slaybaugh, J. C.
"Candidate Functions for Apollo Lunar Surface Television Systems"
February 6 1970

Slayton, Donald K.
"Apollo on Board TV" memorandum
January 20 1969
Record #36436, Program Apollo, Location 070-62, Johnson Space Center History Collection at University of Houston-Clear Lake.

Slayton, Donald K.
"Lunar EVA _not_ based on antenna availability" memorandum
September 12 1968
Report number CF32-8M-207
Record 34923, Program Apollo, Location 079-42/43- A070-34,
Johnson Space Center History Collection at University of Houston-Clear Lake.

Soltoff, Bert M.
"Apollo 15 and 16 Ground-Commanded Television Assembly"
SMPTE Journal vol 81, Dec 1972

Spacecraftfilms
"Project Mercury: A New Frontier"
From the Spacecraftfilms DVD set
December 10, 2005

Spacecraftfilms
"Apollo 1"
From the Spacecraftfilms DVD set
March 20, 2007

Spacecraftfilms
"Apollo 7: Shakedown Cruise"
From the Spacecraftfilms DVD set
February 1, 2005

Spacecraftfilms
"Apollo 8: Leaving the Cradle"
20th Century Fox
From the Spacecraftfilms DVD set
August 19, 2003

Spacecraftfilms
"Apollo 9: Spider Takes Flight"
From the Spacecraftfilms DVD set
February 28, 2004

Spacecraftfilms
"Apollo 10: The Dress Rehearsal"
From the Spacecraftfilms DVD set
January 15, 2005

Spacecraftfilms
"Apollo 11: Men on the Moon"
20th Century Fox
From the Spacecraftfilms DVD set
August 19, 2003

Spacecraftfilms
"Apollo 12: Ocean of Storms"
From the Spacecraftfilms DVD set
May 25, 2005

Spacecraftfilms
"Apollo 13: The Real Story"
From the Spacecraftfilms DVD set

Spacecraftfilms
"Apollo 14: To Fra Mauro"
From the Spacecraftfilms DVD set
August 29, 2006

Spacecraftfilms
"Apollo 15: Man Must Explore"
20th Century Fox
From the Spacecraftfilms DVD set
April 13, 2004

Spacecraftfilms
"Apollo 16: Journey to Descartes"
From the Spacecraftfilms DVD set
June 15, 2005

Spacecraftfilms
"Apollo 17 End of the Beginning"
From the Spacecraftfilms DVD set
November 1, 2004

Spacecraftfilms
"Mission to the Moon"
From the Spacecraftfilms DVD set
December 23, 2005

Space Task Group
"Project Apollo Preliminary Concept of Apollo Spacecraft Communications and Tracking Equipment."
Langley Field, Virginia
March 20, 1961

Space Task Group
"Technical Liason Group Meetings - Selected Excerps"
Langley Field, Virginia
February 14, 1961

Stafford, Tom and Cassutt Michael "We Have Capture"
Smithsonian 2002
ISBN-10: 1588340708 ISBN-13: 978-1588340702

Stanwood, Robert K.
"Television System Evaluation of Proposals" memorandumranda.
February 19, 1962 and April 3, 1962

Stanwood, R.K.
"Preliminary Study of a Digital Television System for use on the Apollo Spacecraft"
March 20 1961

Stoney, William E.
"Definition of Navigation, Location and TV Requirement Systems for Apollo 16 and Subsequent Missions"
October 24 1969

Stoney, William. E.
"Lunar Surface TV for Apollo 16 and Subsequent Missions"
November 6 1969

TePoel, Harold E.
"Report of Trip to GAEC" memorandum
December 7 1964
Record #22932, Program Apollo, Location 064-65,

Johnson Space Center History Collection at University of Houston-Clear Lake.

Truszynski, G. M.
"Post Liftoff TV Coverage for Apollo 15" memorandum
September 7 1971
Record 34926, Program Apollo, Location 073-42

Johnson Space Center History Collection at University of Houston-Clear Lake.

TRW
"Lunar Multipath Signal Characteristics"
Report number N69-40569
October 15 1969

Van der Capellen, A. G.
"Progress Report on TV Transmitter-Exciters and Power Amplifiers"
Report number N65-21334
December 1 1964

Varson, W. P.
"Functional Description of Unified S-Band System and Integration Into the Manned Space Flight Network"

Vavra, Paul H.
"Data to Support Apollo TV Presentation in the May Management Council Meeting" memorandum
May 18 1966
Record #27204, Program Apollo, Location 066-66
Johnson Space Center History Collection at University of Houston-Clear Lake.

Vavra, Paul H.
"Clarification of Data on LEM TV" memorandum
May 20 1966
Record #27221, Program Apollo, Location 067-11
Johnson Space Center History Collection at University of Houston-Clear Lake.

WEC
Engineer's Trip Report
"To watch the Apollo Color Television Presentation from Goddard to Houston."
April 28, 1969
Memo courtesy of Stan Lebar

WEC
"Westinghouse Apollo TV...Faster...Zoom...and Color!"
November 1969
1969

WEC
"Apollo Color Television Subsystem Operation and Training Manual"
June 1 1971

WEC
"Sequential Color Filter System for Television"
Report number 7560
March 6 1968
Report courtesy of Stan Lebar

WEC
"Apollo 14 Television Cameras"
Report number 814-575N
1971

WEC
"Quality Program Plan"
Report number NAS 9-3548
December 18 1964

WEC
"Sun Proof Color TV Camera for Apollo 14 Moonwalk"
1971

WEC
"Westinghouse Builds Color TV Camera and 'mini' Monitor for Apollo 10"
May 16 1969

WEC
"Apollo Lunar Television Camera Operations Manual"
Report number 8418A
August 30 1968

WEC
"Westinghouse TV Cameras bring Apollo Video From Liftoff to Lunar Landscape" memorandum
1970

WEC
"Display Processor for Improved Apollo TV Camera: Final Report"
Report number N69-40571
October 10 1969

WEC
"Lunar TV Final Report: Volume 1"
Report number NAS 9-3548
June 1 1967

WEC
"Westinghouse Engineer"
Westinghouse Report on the lunar TV camera
March 1968

Westmoreland, Perry
"Block II CSM and LEM TV Camera Requirements"
memorandum
September 16 1965

Wilford, John Noble
"Apollo's TV Camera to Be Tested in Flight"
New York Times
January 17, 1966

Wood, Bill
"Apollo Television"
2006
http://www.hq.nasa.gov/alsj/frame.html

Woods, David and O'Brien, Frank
Apollo 8 Flight Journal
[http://history.nasa.gov/ap08fj/]

Wynn, W. D.
"The use of Emphasis in the Apollo CSM and LM USB TV Modes"
Bellcom report number TM 67-2034-6
December 27 1967

Young, Anthony H.
"Lunar and Planetary Rovers: The Wheels of Apollo and the Quest for Mars"
Springer
2006

About the Author

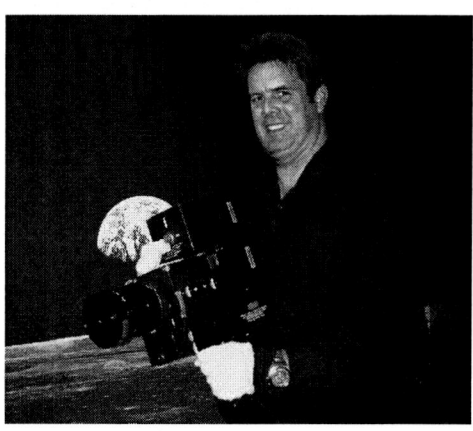

Dwight Steven-Boniecki was born in Sydney, Australia in 1969 a few months before man walked on the moon. He spent much of his childhood fascinated with space exploration – growing up in the shadow of Apollo and under the direct influence of science fiction films such as Star Wars. The latter shaping his desire to work in the film/television industry. After studying television theory at North Sydney Technical College he moved to San Diego, USA and interned at Daniels Cablevision. He returned to Australia and worked at TCN-9 before heading back to university where he majored in Psychology. Following his studies he decided that television was where he truly wanted to be and returned to the industry working for Foxtel, Australia. From there he heard about the expansion on satellite TV in Eastern Europe and jumped on a plane to work in Europe: first in Great Britain for Wizja TV, and then in Germany for CBC / RTL – where he still works today as a transmission controller.

All the while, his interest in space exploration never left him. The advance of DVDs and the internet saw him revisiting the missions he recalled watching as a young child. While watching the missions again, he began to wonder about the technology behind the images he was watching, and so he began researching the television systems developed by NASA mainly to satisfy his own curiosity. To his dismay he discovered that while the information was available, it was not easy to access, and had never been comprehensively written about. He set about changing that, and ended up writing his first book, "Live TV From the Moon". Along the way he befriended many of the people who were directly involved in building the TV cameras which transmitted arguably the most important television signals ever received on planet earth – and is proud to have been able to tell their story.

About the DVD

Included on the accompanying DVD are the following films:

- The 2009 NASA Webcast about the missing Apollo 11 telemetry tapes and the restoration project of the Apollo 11 lunar television footage.

- A slide show of the TV camera hardware accompanied by an interview with Lunar TV tech Paul Coan.

- A slide show of the still images included in this book accompanied by an interview with Lunar TV tech Olin Graham.

- The three Apollo lunar launches as seen by the camera left on the lunar rovers. (Apollos 15, 16 and 17)

- A rare Westinghouse documentary about the construction of the lunar television camera. (Courtesy Stan Lebar)